SLEEPING, DREAMING, AND DYING

SLEEPING, DREAMING, AND DYING

Foreword by

H.H. THE FOURTEENTH DALAI LAMA

Narrated and edited by

Francisco J. Varela, Ph.D.

With contributions by

Jerome Engel, Jr., M.D., Ph.D., Jayne Gackenbach, Ph.D.,
Joan Halifax, Ph.D., Joyce McDougall, D.Ed.,
and Charles Taylor, Ph.D.

Translations by

B. Alan Wallace and Thupten Jinpa

Wisdom Publications • Boston

WISDOM PUBLICATIONS
361 NEWBURY STREET
BOSTON, MASSACHUSETTS 02115

Library of Congress Catalogue-in-Publication Data

Bstan-'dzin-rgya-mtsho, Dalai Lama XIV, 1935-
Sleeping, dreaming, and dying: an exploration of consciousness with the Dalai Lama; foreword by H.H. the Fourteenth Dalai Lama; narrated and edited by Francisco J. Varela; translations by B. Alan Wallace and Thupten Jinpa.
p. cm.
ISBN 0-86171-123-8 (alk. paper)
1. Consciousness--Religious aspects--Buddhism. 2. Sleep--Religious aspects--Buddhism. 3. Dream--Religious aspects--Buddhism. 4. Death--Religious Aspects--Buddhism. 5. Buddhism--Psychology I. Varela, Francisco J. II. Wallace, B. Alan. III. Thupten Jinpa. IV. Title
BQ4570.P75B77 1997
294.3' 422--dc21 97-2448
CIP

ISBN 0-86171-123-8

02 01 00 99 98
6 5 4

Cover Art: *Woman Under Grapes,* by Andrew Stevovich, courtesy of Adelson Galleries, Inc. New York, NY.

Cover design: L.J.SAWLIT

Wisdom Publications' books are printed on acid-free paper and meet the guidelines for the permanence and durability of the Committee on Production Guidelines for Book Longevity of the Council on Library Resources.

Printed in the United States of America.

Table of Contents

Publisher's Acknowledgment

WISDOM PUBLICATIONS GRATEFULLY ACKNOWLEDGES the generosity of the Gere Foundation in funding the production of this book. For many years Richard Gere has been instrumental in spreading His Holiness the Dalai Lama's teachings and message of universal responsibility, compassion, and peace through his own public expressions and by helping Wisdom publish a series of important books by His Holiness. Richard has also steadfastly supported His Holiness's efforts to focus the world's attention on the suffering of the Tibetan people and the threat to their land and cultural heritage posed by the Chinese occupation. We laud his efforts and appreciate his support.

Editor's Acknowledgments

I WOULD LIKE TO EXPRESS my heartfelt gratitude to the many individuals who made the Fourth Mind and Life Conference and this book possible. First and foremost to His Holiness the Dalai Lama for his continued interest and his warm hospitality for these events. To Tenzin Geyche and the Private Office of His Holiness the Dalai Lama, who have been enormously helpful. To Adam Engle, our fearless organizer and chairman of the Mind and Life Institute. To Alan Wallace, whose ideas and support were essential. To Ngari Rinpoche and Rinchen Khandro, both gracious hosts at Kashmir Cottage in Dharamsala and steady supporters of these events. To the invited speakers, who cheerfully leaped into this adventure and whose collective minds' work this book represents. Our generous sponsors Barry and Connie Hershey and Branco Weiss provided the means to transform this vision into reality.

The final edition and transcription was accomplished thanks to the generous work of Phonicia Vuong and Zara Houshmand. Alan Wallace corrected and completed His Holiness's interventions directly from the orginal Tibetan. Tim McNeill, John Dunne, and Sara McClintock at Wisdom Publications are responsible for bringing this book into its final form. Their competence and friendliness made the final stages of publication a smooth, enjoyable conclusion to a long journey.

THE DALAI LAMA

Foreword

WE LIVE IN AN ERA in which science and technology have had a tremendous impact on all our lives. Science, a great product of the human intellect, and the wonderful tool of technology are expressions of our greatest gift—human creativity. Some of their effects, such as developments in communications and health care, have been wonderfully fruitful. Others, like sophisticated weapons systems, have been unbelievably destructive.

Many people have believed that science and technology could solve all our problems. Lately, however, we have witnessed a change in attitude. It has become clear that external progress alone cannot bring mental peace. People have begun to pay greater attention to inner science, the path of mental investigation and development. Through our own experience we have arrived at a point where there is a new awareness of the importance and value of inner mental qualities. Therefore, the explanations of the mind and its workings presented by the ancient scholars of India and Tibet are becoming increasingly valuable in our time. The strength of these traditions is related to developing mental peace. Science and technology are related to material progress. But a combination of these two can provide the complete conditions for obtaining real human happiness.

The series of meetings that we have called "Mind and Life" have been going on for several years. I consider them to be of crucial importance. It is not so long ago that many people viewed common science's objective knowledge and the subjective understanding of inner science as mutually exclusive. In the Mind and Life meetings experts from both these fields of investigation have come together to exchange their experience and different points of view on topics of common interest. It has been a pleasure to discover the great extent to which we have been able to enrich each other's understanding. Moreover, our meetings have been marked not merely by polite curiosity, but also by a warm spirit of openness and friendship.

On the occasion reported in this book we met to discuss sleeping, dreaming, and dying. These are topics that absorb scientists and meditators alike but are also universal elements of human experience. We all sleep. Whether we acknowledge it or not we all dream. And certainly every single one of us will die. Although these issues affect us all, they retain a sense of mystery and fascination. Therefore, I am sure that many readers will be delighted to be able to share in the fruit of our discussions. It remains only for me to express my gratitude to everyone who has contributed to these meetings so far and to repeat my earnest hope that they will continue to take place in the future.

March 25, 1996

A Prelude to the Journey

ALWAYS AND EVERYWHERE, humans have faced two major life passages in which our habitual mind seems to dissolve and enter a radically different realm. The first passage is sleep, humanity's constant companion, transitory and filled with the dream life that has enchanted cultures from the beginning of history. The second is death, the grand and gaping enigma, the final event that organizes so much of individual existence and cultural ritual. These are ego's shadow zones, where Western science is often ill at ease, far from its familiar territory of the physical universe or physiological causality. In contrast, the Tibetan Buddhist tradition is fully at home here; in fact, it has accumulated remarkable knowledge in this area.

This book is an account of a week-long exploration of these two great realms of radical transformation of the human body and mind. The exploration takes the form of a unique exchange between the Dalai Lama, with a few of his colleagues in the Tibetan tradition, and representatives of Western science and humanism. The exchange was the fourth in a series of biennial meetings called Mind and Life Conferences. It was a private, highly structured dialogue, that took place over five consecutive days in October 1992 in Dharamsala, India.

On Monday morning, all the participants gathered in the Dalai Lama's living room to begin our journey. His Holiness the Dalai Lama appeared promptly at nine o'clock, as was his habit. He entered, beamed at everyone, and invited us to sit down. The speakers were grouped in an inner circle of comfortable couches, with observers and advisers in an outer circle. The atmosphere was relaxed and informal: no television cameras, no high podium, no formal speeches. The unique magic of the Mind and Life Conferences was being created once again.

The Dalai Lama opened with some friendly words. "Welcome to all of you! There are many old friends among you, and perhaps you have the feeling that coming to Dharamsala is like coming home. I am very happy to have another Mind and Life Conference. I believe our previous conferences were of great benefit, at least to me and to people interested in these issues."

He then turned to a more global perspective. "Since our last conference there have been many changes on this planet. One of the most important is the disappearance of the Berlin Wall. The threat of a nuclear holocaust is now more or less gone. Although problems remain, the world is now more favorable for genuine, lasting peace. Of course, killing continues here and there, but overall the situation has improved. Everywhere people are talking about democracy and freedom. That also is of great significance. I believe that the desire for happiness is an essential part of human nature. Happiness comes from freedom. On the contrary, dictatorship of any kind is very harmful for the development of the community. In the old days, certain people had some enthusiasm for authoritarian regimes, but nowadays this has changed. The younger generation is devoted to freedom and democracy. We may change the world, at least in terms of social inequalities. The strength of the human spirit again has the upper hand."

His Holiness went on to set the context for our meeting. "Now we have these two fields, science and spirituality, in which we are supposedly involved," and as he said this he laughed wholeheartedly and contagiously. That laughter was to be as present in the days to come as was the probing intelligence of all the participants, and the group was never far from a sense of humor. "It seems that scientific research reaches deeper and deeper. But it also seems that more and more people, at least scientists, are beginning to realize that the spiritual factor is important. I say 'spiritual' without meaning any particular religion or faith, just simple warmhearted compassion, human affection, and gentleness. It is as if such warmhearted people are a bit more humble, a little bit more content. I consider spiritual values primary, and religion secondary. As I see it, the various religions strengthen these basic human qualities. As a practitioner of Buddhism, my practice of compassion and my practice of Buddhism are actually one and the same. But the practice of compassion does not require religious devotion or religious faith; it can be independent from the practice of religion. Therefore, the ultimate source of happiness for human society very much depends on the human spirit, on spiritual values. If we do not combine science and these basic human values, then scientific knowledge may sometimes create troubles, even disaster. I think the achievements of science and technology, for all their awful destructive powers, are immense. But because they bring us fear, suffering, and anxiety, some people consider them to be negative.

"Scientific knowledge can be seen as a faculty of human intelligence—it can be used either positively or negatively, but in itself it is morally neutral. Whether it becomes beneficial or harmful depends on one's motivation. With proper motivation scientific knowledge becomes constructive. But if the motivation is negative, then the knowledge becomes destructive. These conferences will eventually demonstrate ways for science and spirituality to work together more closely. I think each of us has already made some contribution in this respect, and I'm quite sure this conference will as well. We may contribute something, and if not, at least there will be no harm." This sentence was followed by a good laugh from everyone. His Holiness concluded with a beaming smile, "So that's good. For these reasons, with these feelings, I welcome you all to my home."

It was my turn, as chairman and scientific coordinator, to reply to his welcoming words. By then, it was easy to say that we were all quite moved to be there and to have the opportunity to be part of this singular adventure.

Charting Ego's Shadow Zones

I went on briefly to set the stage for the week's agenda. Basically, we would focus on areas of mind that are essential for human existence, yet difficult for Westerners to understand: sleeping, dreaming, and dying. In keeping with the spirit of these meetings, we wanted to address these topics in the widest possible sense, so that broad surveys of what is happening in the West would be presented by researchers involved in their respective fields. The first three days would be devoted to sleep and dreaming, and the last two days to dying. I will briefly describe the reasons for these thematic choices, and introduce the invited speakers. Detailed biographical sketches of participants can be found at the end of this book.

The first day on the topic of sleep and dreams was devoted to neuroscience, which studies the brain's involvement in sleep as a biological process. It was essential to have on hand some basic results of one of neuroscience's most active fields: sleep research. This first morning presentation had been entrusted to a specialist in the field, Michael Chase (University of California at Los Angeles), who had to cancel at the last minute. Fortunately we had a distinguished group of neurobiologists present in Dharamsala: Clifford Saron (University of California at San Francisco), Richard Davidson (University of Wisconsin at Madison), Gregory Simpson (Albert Einstein School of Medicine), Robert Livingstone (University of California at San Diego), and myself (Centre Nationale de la Recherche Scientifique, Paris). Collectively we prepared a presentation on basic sleep mechanisms, and it was decided that I should present it to His Holiness.

The second day would address dream work in psychoanalysis, which is somewhere between a scientific psychology and a humanist practice. It is a tradition that has left a deep mark on Western views of the structure of mind and the role of dreams. Although some readers might prefer that another psychotherapeutic school had taken its place at the conference, it seemed to me that the Freudian tradition was the most influential and pervasive. The point was not to champion contemporary Freudian schools, but to bring to the discussion a sense of how dream work has become part of Western thinking and culture. Joyce McDougall, a well-known and respected figure in contemporary psychoanalysis in both

Europe and in the larger English-speaking world, was chosen as the presenter.

During the third day we would move to a more recent and controversial area within the study of dreams: the phenomenon of lucid dreaming. We chose this topic because on the one hand it has received some scientific attention in the West, and on the other hand it has been a very active field of study in the Buddhist tradition. We hoped that some connections with Tibetan Buddhism would emerge. Jayne Gackenbach, a psychologist at the University of Alberta who had been active in this field for some years, would be the presenter.

On the fourth and fifth days we were to cover the issue of dying. We reduced this enormous topic to two main themes. On the fourth day we wanted to cover the biomedical understanding of the process of dying. Although medicine pervades our lives, once a person is considered over the threshold, the entire observational and experimental machine of modern biomedicine grinds to a halt. Little is known about death's intimate, final stages. We called on Jerome ("Pete") Engel to fulfill this difficult task. As a member of a large biomedical facility at the University of California at Los Angeles, and a world-renowned neurologist, he seemed better prepared than other professionals to explore this uncharted ground.

Finally, we would close on the fifth day with our second death-related theme, a view of recent research on how humans have traditionally grappled with death through so-called near-death experiences. These were, again, controversial waters for established science, but they are areas that are clearly evoking a huge interest in the West. We hoped to find links between this research and one of the most original areas of experiential and philosophical importance in the Tibetan tradition, the human encounter with death. As the day's speaker, we chose Joan Halifax, a medical anthropologist who had been a pioneer in this field in the sixties and seventies, and had extended her observations to shamanic traditions.

That was, in a nutshell, the agenda of the meeting in regard to its scientific content. However, as in the previous Mind and Life

Conferences, we found it essential to include an overview of the philosophical underpinnings of the Western perspective on these topics. This was crucial, though it might at first glance seem surprising. Clarifying the conceptual basis of a discipline or a history of ideas lays out fertile terrain on which to build discussion. The Tibetans, masters in the art of conceptual clarity, were always very receptive to this dimension of our previous discussions. We had asked Charles Taylor from McGill University to fulfill this role now, since he was known for his perceptive studies on the modern self and its historical roots.

Cross-Cultural Dialogue and the Mind and Life Conferences

Before we begin our journey with Charles Taylor's exploration of the concept of self, let us pause for an account of the background that led to this unique gathering. As I mentioned above, this conference is the fourth in a series of similar meetings, starting in 1987, that came to be called the Mind and Life Conferences. The rich dialogue that fills this book shows that this fourth conference was a resounding success. This was not a matter of mere chance. Intercultural exchanges are notoriously difficult to stage properly, for they easily slip into the pitfalls of superficial formality or hasty conclusions. To give some idea of how we avoided those pitfalls, I will briefly describe our approach to the process of dialogue. And since the exchanges in the previous Mind and Life Conferences were an integral part of the dialogue that unfolded in this fourth meeting, I will also sketch the content of those conferences. A more extensive account of the origins of the Mind and Life Conferences and information about the participants can be found in the Appendix.

As with all such endeavors, the Mind and Life Conferences began merely as an intriguing notion shared among a few friends and colleagues. I had been interested since 1978 in the intercultural and interdisciplinary bridges that can enrich modern science (particularly the neurosciences, my specialty). It was not until 1985, however, that an opportunity to act on these interests presented itself. In that year, Adam Engle and I began to plan a dialogue between Western scientists and His Holiness the Dalai Lama, one of the most accomplished practitioners and theorists within contemporary Buddhist traditions. Two more years of organizational groundwork were to pass before the first Mind and Life Conference finally took place.

For the conferences to succeed, we learned that the scientists chosen need not necessarily be famous names. Of course, they needed to be competent and accomplished within their own fields, but they also had to be open-minded—and preferably not too ignorant of Buddhism. We adjusted the agenda as further conversations with His Holiness clarified how much of the scientific background we would need to fill in. In the end, His Holiness agreed to set aside a full week for us, a measure of the importance he placed on these discussions. In October 1987, the first Mind and Life Conference took place in Dharamsala. It covered the basic ground of modern cognitive science, the most natural starting point for contact between the Buddhist tradition and modern science. Many of the basic features of the meeting would be maintained and refined at subsequent Mind and Life Conferences.

One important feature ensured that the meetings would be fully participatory. We created a format that called for presentations by Western scientists each morning, with the afternoons devoted solely to discussion. In this way, His Holiness could be briefed on the topic at hand. We insisted that this presentation be made from a broad, nonpartisan point of view for fairness. The presenter could spell out his or her own preferences and judgments freely in the afternoon.

A second important issue was translation. We were able to secure the services of many wonderful interpreters, and at every session two were present, one on each side of the Dalai Lama. This allowed quick, on-the-spot clarification of terms, which is absolutely essential for moving beyond the initial misunderstandings that sometimes arise in dialogues between two vastly different traditions.

A third key aspect of the meeting was that it was entirely private: no press, no television cameras, and very few invited guests. This stood in sharp contrast to meetings in the West, where the public image of the Dalai Lama makes relaxed, spontaneous discussion increasingly impossible. Thus, meeting in Dharamsala allowed us a kind of protective freedom to conduct our exploration.

The agenda for the first conference introduced broad themes from cognitive science: scientific method, neurobiology, cognitive psychology, artificial intelligence, brain development, and evolution.[1] The event was an enormous success in that both His Holiness and we felt that there was a true meeting of minds, and some substantial advances in bridging the gap between Western and Buddhist thinking. The Dalai Lama encouraged us to continue with further dialogues on a two-year basis, a request that we were only too happy to honor.

Mind and Life II took place in October 1989 in Newport, California. It was a two-day event that focused more specifically on neuroscience. The event was especially memorable, as we learned on the first morning that His Holiness had been awarded the Nobel Prize. The third Mind and Life Conference dealt with the relationship between emotions and health.[2] At its close, His Holiness again agreed to continue the dialogue in a subsequent meeting—that meeting constitutes the adventure reported in this book.

It was against this background that we met for the fourth conference, with a sense that our efforts were beginning to bear fruit. And now, once again in Dharamsala for a week, we would push the dialogue deeper, into the territory of sleeping, dreaming, and dying. Sitting with me were the contributors whose voices the reader will hear in this book. As in the past, Thupten Jinpa and Alan Wallace would serve as our very able translators.

It seemed best to begin with a learned philosopher's account of the Western conception of what it is to be a self. Hence, I asked Charles Taylor to be the first to take the "hot seat," the armchair next to His Holiness that each participant in turn would occupy in the days to follow.

1

What's in a Self?

A History of the Concept of Self

PAST CONFERENCES WITH HIS HOLINESS in the Mind and Life series had taught us that having a professional philosopher conversant with the scientific topic at hand was very useful. One of the main reasons is that in the Tibetan tradition philosophical reflection and discipline are highly valued and cultivated. A Western philosopher among scientists often provided valuable bridges and alternative formulations that were clearer and closer to the Tibetan tradition. For the topic of this conference, Charles Taylor, a well-known philosopher and writer, was an ideal choice. In his recent book *Sources of the Self*, he had drawn a vivid and insightful picture of how we in the West have come to think about the thing we call the self.[3] He launched into the subject with speed and precision.

"I'd like to talk about some of the most important aspects of the Western understanding of the self. To do that I'd like to paint a very broad picture of the concept's historical development. I think a good place to start would be with the very expression *the self*. In our history it's something quite new in the last couple of centuries to say 'I am a self.' Before this, we never used the reflexive pronoun *self* with a definite or indefinite article (such as *the* or *a*). The ancient Greeks, the Romans, and people of the Middle Ages never treated it as a descriptive expression. We could say today that there are thirty selves in the room, but our ancestors wouldn't have said that. They would have perhaps said there are thirty *souls* in the room or employed some other description, but they wouldn't have used the word *self*. I think this reflects something fundamental in our understanding of the human agent, something very deeply embedded in Western culture.

"In the past one would have used the words *myself* or *I* indistinctly, but the word *self* is now used to describe what a human being is. I would never describe myself as 'I.' I just use that word to refer to myself. I would say: What am I? I'm a human being; I'm from Canada. I describe myself in that way, but in the twentieth century I might say 'I am a self.' The reason I think that's important is because we choose the descriptive expressions that reflect what we think is spiritually or morally important about human beings. That's why our ancestors spoke of us as souls; that's what was spiritually and morally important to them.

"Why did people become uncomfortable with that usage and why did they shift over to using *the self*? Part of the story is that they found something spiritually significant in describing us as selves. Certain capacities that we possess to reflect on ourselves and operate on ourselves became morally and spiritually central to Western human life in a crucial way. Historically, we sometimes called ourselves 'souls' or 'intelligences,' because those concepts were very important. Now we speak of ourselves as 'selves' because there are two forms of concentration and reflection on the self which have become absolutely central to our culture, and which are also in tension with each other in modern Western life: self-control and self-exploration.

"Let's first look at self-control. Plato, the great philosopher of the fourth century B.C.E., spoke of self-mastery. What Plato meant was that one's reason was in control of one's desires. If one's desires were in control, one would not be master of oneself."

"Very wise!" the Dalai Lama interjected.

"But interestingly, self-control had a very different meaning for Plato than it has in the modern world. For Plato, reason was the capacity in human beings to grasp the order of the universe, the order of the 'ideas,' as he called them, that gave shape to the universe. To have reason commanding one's soul was the same as having the order of the universe commanding one's soul. If I look at the order of things, my soul comes into order from love of that order. So it was really not control by myself as an agent alone; it was control by the order of the universe. Human beings were not encouraged to reflect inward on the contents of their own souls, but rather to turn outward to the order of things.

"Christianity changed that very profoundly with Saint Augustine in the fourth century C.E. He was influenced by Plato, but he had a very different view. His idea was that we can get close to God by turning inward and coming to examine what we have within ourselves. We discover that at the very heart of things we depend upon the power of God, so we discover the power of God by examining our selves.

"So we had these two spiritual directions: one, Plato, turning outward and the other, Augustine, turning inward, but still with the intention of reaching something beyond ourselves, which is God. A third change comes in the modern West. Take the seventeenth-century philosopher Descartes as an example. Descartes believed in God and he thought of himself as following Augustine, but he understood something quite different by the idea of self-control: the instrumental control that I as an agent can exercise over my own thinking and over my own feeling. I stand in relation to myself as I stand to some instrument that I can use for whatever purpose I want. Descartes reinterpreted human life as the way we concentrate on our selves as instruments. We came to see our bodily existence as a mechanism we can use, and this happened in the great age when a mechanistic construct of the universe arose.

"The modern idea of self-control is very different from Plato, because the order of the universe is no longer important or relevant. It's not in control. I am no longer even turning inward to get beyond myself to God; instead I have a self-enclosed capacity to order my own thoughts and my own life, to use reason as an instrument to control and order my own life. It becomes very important for me to order my own thinking, to keep it operating in the right way by the right steps, to relate to it as an object domain that I can somehow dominate. This has become absolutely central to Western life. It's one way we begin to think of ourselves as 'selves,' because what's really important is not the particular content of our feelings or thinking but the power to control it reflexively."

As was customary in our Mind and Life meetings, presentations were peppered with clarifying questions from the Dalai Lama. In fact, by following the type of questions asked, the reader can gain accurate insight into the gaps between the Tibetan and Western traditions. In this case, he politely interrupted Charles: "Would you say that this self as a controller has the same nature as the body and mind that are being controlled? Or is its nature distinct from those of body and mind?"

"For Descartes, it was the same thing," came the reply. "But the self came to be seen as something distinct because it doesn't have any particular content itself. It is just the power to control whatever thought content or bodily content occurs."

Self-Exploration and Modernity

The discussion turned to self-exploration. "At the same time as Descartes was developing these ideas, another important human capacity appeared in the West: self-exploration. This grew out of the flourishing of Christian spirituality that was inspired by Augustine, which led people to turn to self-examination, examining their souls and examining their lives. Self-examination, too, developed beyond the original Christian form, and in just the last two hundred years it has become an extraordinarily powerful idea, which is now fundamental in the West, that each human being has their own particular, original way of being human.

"There were ancient practices of self-exploration, but they always started from the assumption that we already know what human nature is, and our task is to discover within ourselves what we already know to be true. In the last two hundred years, the assumption is that we know in general what human nature is, but because every human being has their own particular, original way of being human, we therefore have to draw that nature out of ourselves by self-exploration. This has opened a whole range of human capacities which are thought to be very important. How do you explore yourself? You find what is not yet said, what is not yet expressed, and then find a way of bringing it to expression. Self-expression becomes very significant.

"How do you find the languages of self-expression? In the West in the last two hundred years it's been thought that people can find the best languages of self-expression in art, whether poetry, visual art, or music. It is a feature of modern Western culture that art has an almost religious significance. In particular, people who have no traditional religious consciousness often have this deep reverence for art. Some of the great performers in the West have an aura around them—famous, beloved, and admired—that is unprecedented in human history.

"So we have these two practices of self-relation: self-control and self-exploration. Because they are both crucially important, we have come to think of ourselves as 'selves' and to refer to ourselves that way without reflecting. Both practices belong to the same culture but they are also profoundly at odds, and our civilization is constantly battling itself over this. You see it everywhere you look.

"You see it in the conflict today in the West between people with a very strict, narrow, technological orientation to the world and themselves, and those who oppose them in the name of ecological health and openness to oneself because the technological stance of self-control also closes off self-exploration.

"You get it in attitudes to language. On one side, language is conceived as a pure instrument controlled by the mind, and on the other side are conceptions of language that have led to some of the richest discoveries about human understanding—language as the house of being, language as what opens up the very mystery of the human being.

"What draws self-control and self-exploration together is that they have a common source: a conception of the human being that focuses on the human being in a self-enclosed way. Plato could not grasp the human being outside of the relationship to the cosmos, and Augustine couldn't grasp the human being outside the relationship to God. But now we have a picture of the human being in which you may also believe in God, you may also want to relate to the cosmos, but you can grasp the human being in a self-enclosed fashion with these two capacities of self-control and self-exploration. It also has meant that perhaps the most central value in the moral and political life of the West is freedom, the freedom to be in control or the freedom to understand who one is and to be one's real self."

Once more the Dalai Lama clarified a key issue: "Is there an underlying assumption that self-control necessarily implies a self-existent or autonomous self, whereas self-exploration implies that that's doubtful?" Charles answered that that was not necessarily the case, that self-exploration also presupposes a self, but opens the possibility that the exploration can go beyond that. The stance of self-control assumes that there is a controlling agency and never calls that into doubt. For instance, Descartes' philosophy famously starts with the certainty that I, myself, exist. The entire edifice of scientific understanding of the world is built on that certainty.

Science and the Self

After painting this masterful picture of what it is to be a modern self, Charles brought the discussion back to the task at hand by relating these concepts of the self to the scientific tradition, and in particular to certain modes of scientific understanding that had already figured in earlier Mind and Life Conferences. "Take, for instance, the type of cognitive psychology that understands human thinking on the model of the digital computer. This is an extraordinary idea, a crazy idea for some of us, I have to admit, but with immense imaginative power.

"Going back to Descartes himself, the stance towards the self as a domain of instrumentality views the self as a kind of mechanism. The idea that we are, at bottom, just a mechanism is very congenial to this field. At the same time, Descartes put a tremendous emphasis on clear, calculative thinking. In other words, thinking would be clearest when it followed certain formal rules where you could be absolutely certain that each stage was a valid step from which to proceed to the next valid step. The wonderful thing about computers is that they combine this absolutely formal thinking with a mechanistic embodiment. People who are deeply moved by this side of Western culture are endlessly fascinated by computers and therefore are ready to make them the basis of their model of the human mind.

"On the other side are human sciences that grow out of the long tradition of self-exploration. One of the changes that has occurred in language in the West, along with using words like the self, is the development of a very rich language of inward exploration. Expressions like 'inner depths' are very much part of our culture—I would love to know if something similar exists in Tibetan. The idea is that each of us has to carry out a very long and deep exploration in ourselves; we think of that which we don't fully understand as somewhere deep down and we think of these depths as inner. This emerges in another strand of Western scientific discourse, an example of which is psychoanalysis.

"Another direction of self-understanding that belongs to the line of self-exploration in the West today is *identity*. This is another word that is used today in a quite unprecedented sense. We often talk about discovering 'my identity' or we talk about our teenagers having a crisis of identity—of not knowing their identity, and the pain and drama of discovering it. My identity is who I am. In a sense, this is a way of describing myself as a spiritual being because when people talk about what they think their identity is, they're really talking about the horizon from which they understand what is really important to them and what is vital in human life. In other words, the spiritual horizon of each person is understood as being bound by who that person is. Once again this reflects the search for what is particular to each human being. It is in this domain that explorations of new ways of understanding the human being are taking place in the West.

"This is a point that opens some very interesting and illuminating contact between the Western view and the Buddhist view. The discourse of identity allows for the possibility that I can radically rediscover and redescribe who I am; that I can discover that who I thought I was is not really correct and has to be re-understood and redescribed. Moreover, it's in this domain that some Western philosophies have begun to question the very certainty of the self as a circumscribed entity. They have raised questions such as, 'Is there really a unitary self?' This is the area in which exploration is going on, the frontier of uncertainty about the very nature of the self. Part of this philosophical movement is a reaction to the concept of self-control, which always seems very clear about the self as the controlling agency and never doubts its unity. This cultural war has resulted in modes of self-understanding that question whether we are in control, whether there are no deep resources within us that escape the self, and whether therefore self-exploration might not lead to something very different and disconcerting, something new and strange."

The Self and Humanism

The presentation had reached its natural conclusion and a flurry of discussion started among all participants. The next question from the Dalai Lama was a bit less clear and a perfect example of the difficulty in making the leap between one tradition and another vastly different one: "Is there not a special relationship between this strong emphasis on the self and humanism? I have heard two very different connotations of the term *humanism.* On the one hand, there is the very positive sense of humanism as ennobling the self, endowing the self with a certain initiative or power. As a result, the self does not seem so much a pawn of God or any other external agency. In this sense humanism seems to be something positive. On the other hand, humanism in a very different context appears to be negative, with its emphasis on the self and its view of the environment simply as something to be manipulated and exploited by the self. If this is the case, how do these two meanings of humanism fit together, and which is in fact the more prevalent sense of the term?"

Charles replied, "One of the meanings of the word *humanism* has included this concentration on man, on the human being. As I said earlier, the two modes of self-control and self-exploration allow us to draw a circle around the human being and focus on that being. But humanism is also very varied and parts of it are in conflict. The two senses of humanism you have heard are two sides of the same coin in Western development. The original humanism of affirmation was relatively blind to the relationship of human beings to the rest of the cosmos. And there is now indeed a chastened humanism, among other things, the one that has learned wisdom of the self's connection to the cosmos, but it is not the original one."

The Dalai Lama probed further: "When you speak of the cosmos, aren't human beings part of the cosmos rather than something separate? If the cosmos is understood as referring to the external environment and human beings are regarded as individual agents existing inside it or even outside of it, aren't human beings still thought to be products of the natural elements?"

"Yes, but in the view of the modern humanism that placed us as users in relation to instruments, the cosmos surrounding us was something that we could and ought to control. Originally, Descartes and others held a very strong dualism in which the human soul was thought to be quite separate from the cosmos; but later you are absolutely right. Another mode of humanism explains human beings in terms of these natural elements, a very reductive and a very arrogant stance of control. Indeed, I think there is a profound contradiction in this position. But a contradictory position is sometimes lived because it's very deeply embedded in a culture."

"So basically, both persons and the whole cosmos could be included in the word *humanism*. Does the term also imply a denial of the existence of a deity?"

"Not usually, but there are some people who use the word in that way," answered Charles. "In England there is a Humanist Society whose members have in common simply that they are atheists. On the other hand, a great Catholic philosopher of this century wrote a book called *Christian Humanism*."

Non-Self in the West

As tea was brought in, the Dalai Lama pursued his question on the relationship between the self and the cosmos, asking everyone around the table: "Descartes seems to define the soul as being independent from the cosmos in general and from the body in particular. What about the modern sense of the self? Is the self seen as an independent agent and something different from the body? What is its relation to the cosmos at large? Now that the self has been secularized, is it no longer possible to continue to conceive of the self as independent of the cosmos?"

Everybody deferred to Charles: "Logically and metaphysically, it doesn't make sense to conceive of the self as separate, but this is the interesting point here. This whole way of understanding ourselves involves each person as a scientist or agent, taking a controlling stance toward the body and the cosmos. There is an implicit self-understanding that contradicts the explicit doctrine of the science. This is one of the great pragmatic self-contradictions of this metaphysical, materialistic stance in the West. The scientific doctrine says that it's all just mechanism, including the self, but in order to get that doctrine you have to take a stance as a controlling agency toward the world. So this same agent has a sense of almost angelic or even godlike power over the world. There is a split in consciousness that is deeply illogical but existentially very understandable."

The Dalai Lama then asked, "In the modern West, when one thinks 'I' or 'I am,' does this necessarily imply that the 'I' so conceived must be posited as being independent or autonomous?"

Charles's answer was very Buddhist in flavor. "If you ask people, they say no. But in the way they actually live it, the answer is yes, very powerfully, and much more so than our ancestors who thought of themselves more as part of a larger cosmos."

Joan Halifax interjected, "In the evolution of the self was there ever a non-self posited, the idea that in fact human beings didn't have a separate self-identity?"

Charles answered, "There are such phases in Western development. For instance, the medieval Aristotelians thought that the really important part of us, the active intellect, was absolutely a universal thing and not particularized. The famous Islamic philosopher Ibn Rushd Averroës thought that also, but he had great problems with mainstream Islam. It was because of Ibn Rushd that Aristotelianism had problems entering Christendom; it's only when Albert the Great and Thomas Aquinas managed to reintroduce the idea of a personal intellect that it was allowed in."

The discussion went around for a while in this questioning vein as everybody sipped tea. It was then time to change the stage, and to plunge into the first scientific presentation: a view of the brain in sleep and dreaming.

2

Brain's Sleep

Sleep in Neuroscience

I EXCHANGED PLACES WITH Charles Taylor, sat down, and smiled at the Dalai Lama, who was looking at me with piercing eyes. It was not my first time in the hot seat, and yet looking around the room I could not but feel moved by the special nature of the occasion. After a brief silence, I began.

"Your Holiness, after this very clear and vivid introduction to notions of the self, I think it's useful to start off with the neuroscience of sleep. I agree with Professor Taylor that science offers the hope that we can account for the entire mind as a mechanism. But another stream within science also acknowledges that there is something that the current neural mechanisms do not fully explain, and it's usually phrased in terms of either self or consciousness. Consciousness is a 'bag term,' a place where we throw anything we don't yet understand, anything that eludes the idea of mind as a computer or a set of neurological processes. In the language of science, the word consciousness often describes whatever is deepest, in the sense of a depth that we haven't yet reached. It is important here for us to remain aware of this unresolved tension within science.

"Why is this realization particularly relevant in understanding the neuroscience of sleep? All research on sleep inevitably deals with radical changes in one's identity, one's self, or one's consciousness. When you go to sleep, all of a sudden you are not there. This automatically raises the elusive notion of self, from which some neuroscientists would rather shy away, not to mention the whole mysterious issue of dreaming.

Early Ideas

"Let's move on to the specifics of the neuroscience of sleep. Looking historically at the research on sleep, we see that the main discoveries have all disproved the view that sleep is passive. Neuroscience started off with the traditional idea that sleep is like switching off the lights of the house, and that human beings left alone with nothing to do will fall asleep.

"Advances in research very quickly made it clear that sleep is an active phenomenon. It is a state of consciousness with its own laws. It was Sigmund Freud who first articulated that sleep is an active process. Although he started as a neuroscientist, Freud moved in a different direction, toward psychology, about which we will hear later this week. Around 1900, the first researchers tried to define sleep physiologically. Around 1920, a French scientist named Henri Pieron expressed the dominant modern view of sleep as having three characteristics. First, it is a periodic biological necessity. Second, it has its own internally produced rhythm. Third, it is characterized by an absence of motor and sensory functioning.

"I'll skip several other milestones and jump ahead to a discovery that is of great relevance for us here. In 1957, a group of American researchers described what is known today as the state of REM sleep. REM stands for rapid eye movement. This discovery marks the beginning of the mainstream research that continues very actively today.

The Basics of the EEG

"Between 1900 and 1957, neuroscience had come to understand in depth the electrical phenomena of the brain, which made the REM discovery possible. We are going to take a detour now in the presentation, leaving sleep aside for a moment, and speak about recording electricity from the human brain by means of the electroencephalogram (EEG). For half a century, the EEG has been the main noninvasive method of investigating human brain activity. (Recent methods of brain imaging constitute a very important alternative and complement, but let us stay with EEG for the time being.) Without some understanding of the techniques and biology of the EEG, we cannot appreciate the neurobiology of sleep.

"The EEG can record a surface electrical potential only because the cortex (the brain's outer layer) is organized with large pyramidal neurons (nerve cells) aligned regularly next to each other. These pyramidal cells are the main neurons that compose the gray matter on the surface of the brain. They receive signals from other regions of the brain through the axons of the white matter. Axons (nerve fibers) transmit action potentials, rapid electrical signals that affect the neurons at their synapses, or point of junctions. If the electrical activity of the neurons overlaps in time, the electrical potential is large enough to read at the surface, although very faintly (fig. 2.1). The electrical signals measure a few millionths of one volt.

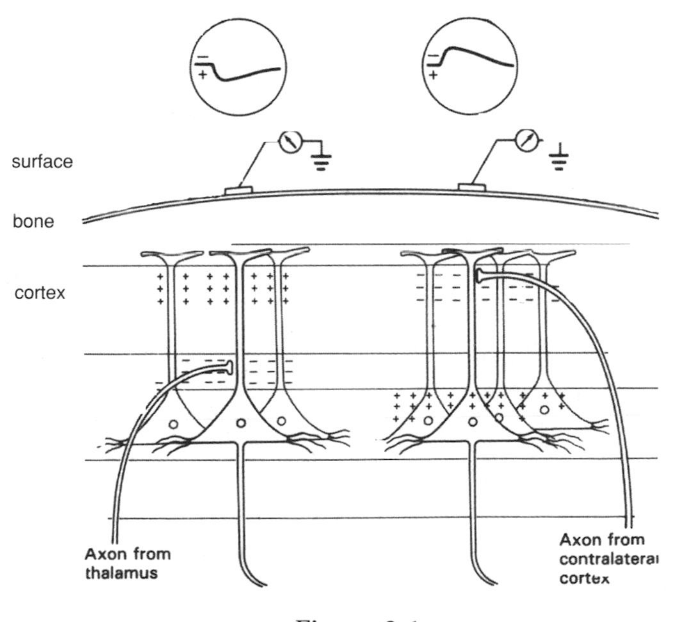

Figure 2.1

Cross section of a portion of the skull bone and underlying cerebral cortex. On the left side of the diagram, the action potential occurs as a negative charge induced by the axon, with a corresponding positive charge in the ascending dendrite (extensions). An electrode placed at the surface shows a wave with a slight deflection. A different incoming axon from another region of the brain induces the opposite electrical pattern. (From Kandel, Schwartz, and Jessel, Principles of Neural Science, *3rd ed., Norwalk, CT: Appleton & Lange, 1991, p. 784. Printed with permission from Appleton & Lange.)*

"What happens if axons with opposite electrical patterns are active at the same time? The result is a flat recording, as the positive contributions are counterbalanced by the negative ones. This desynchronized EEG is like people talking at a cocktail party; a synchronized EEG is like people singing in a choir. In a synchronized EEG, many neurons change from a positive to a negative charge in unison and without counterbalance. The result is a large amplitude wave form generated at the surface.

"Thus, the EEG is the spatial summation of many, many neurons. This summation reflects the underlying pattern of activity near the point of the cortex being recorded. So an EEG recording is a local signal. It is also indirect: many different neuronal patterns give rise to identical surface recordings. Allow me now to stop being abstract and actually present to you the brain waves of our colleague Dr. Simpson!"

We had set it up, of course. As I finished the sentence, Greg Simpson walked into the room fully wired with an electrode cap and dangling wires. Two friends carried in a portable EEG recording machine with a color screen, and placed it on the table in front of the Dalai Lama. Greg sat next to him and was promptly connected to the EEG machine. It all worked smoothly and produced much applause and merriment all around (see fig. 2.2 on colorplate between pages 158 and 159). Once the excitement subsided, I leaned over to His Holiness and explained how the screen showed three active points of recording. The wiring was quite good and it was easy to see the different points of recording, showing a typical, rapidly changing EEG. I asked Greg to close his eyes in order to induce larger amplitude waves in the occipital (back) cortex, in contrast to the smaller amplitude traces that appear when the eyes are open. His Holiness was quite impressed: he could tell that Greg had closed his eyes even without looking at him!

After taking in the whole display, he asked me, "Is there a difference between sitting very quietly in a totally nondiscursive, nonconceptual way, as opposed to holding one single-pointed thought?"

I had to smile. "You have just defined a research project for the next ten years. We don't know the answer to your question because there has not been much interest in studying trained, stable minds." His Holiness then asked what happens to the EEG during speech. I explained that the electrical charge of muscles moving in speech produces a false reading. After a few more questions and clarifications, the machine, having made its point, was removed, and Greg was disconnected. We resumed our places and I continued with the formal presentation.

"As you saw, Your Holiness, in a setup as simple as this, you can distinguish at least the two states of alertness and relaxed wakefulness. Using more electrodes and more complex analyses, researchers can detect and classify many brain patterns, including states of sleep, language behavior, lateralization (right or left brain functioning), and so on."

Sleep Patterns

"Biologists are proud to have discovered that human and animal bodies have many different intrinsic rhythms: hormonal, circadian, temperature control, urination, and many others. They do not necessarily work together, but autonomously. Consider, for instance, the circadian rhythm of night and day. We can study this by keeping people in total darkness in deep caves for two or three months, completely isolated from the rest of the world. Their day and night are no longer related to light from the sun, yet they continue to sleep and wake in cycles that run freely without any external constraints. An adult left in this situation starts to shift away from the rhythm related to the sun, and settles into his own internal rhythm, which varies between subjects. A typical adult has a twenty-five hour rhythm."

His Holiness asked, "Does this rhythm change happen because of the person's thoughts, ideas, and expectations, or is it purely physical in nature?"

I had to smile again. "It's hard to do the experiment on somebody without expectations. There is a lot of variability in the results, which is most certainly influenced by a person's style, but it is a very difficult dimension to explore."

"Is he just lying down all the time?"

"No. When he is awake, he can explore the contents of the cave. He can turn on a small light and do activities such as cooking. There have been extreme cases of people who have gone into complete darkness, just sitting there like prisoners in confinement. All of that accounts for a lot of variation. But the average Western adult has a pattern of 25 hours, with significant variability.

"Until recently the predominant idea was that sleep is simply switching off the machine, letting it cool off, as it were. Modern sleep research began with the discovery that sleep has an intrinsic rhythm. This led to the investigation of the fine patterns in each phase of this cycle, for which the use of the EEG was very important.

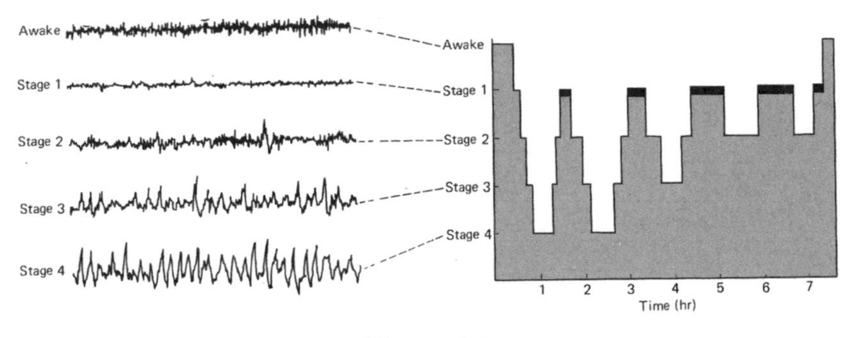

Figure 2.3

The hours of sleep of a normal adult are shown, with sample recordings of the EEG as it changes during various stages. (From Kandel, Schwartz, and Jessel, Principles of Neural Science, *3rd ed., Norwalk, CT: Appleton & Lange, 1991, p. 794. Printed with permission from Appleton & Lange.)*

"We can see from the EEG patterns recorded at different times that not all hours of sleep are the same: there are *stages* within sleep (fig. 2.3). A waking EEG, like that you saw of Dr. Simpson's brain, includes a mixture of rhythms at various frequencies. There are some episodes of higher amplitude waves around 10 hertz (1 Hz = one cycle per second). These are the so-called alpha band brain waves. Traditionally the different ranges of frequencies have been given names with Greek letters; for instance in a normal awake EEG we never see high-amplitude slower waves from the delta band (about 2-4 Hz). When the person falls asleep, this pattern shifts rather dramatically.

"In the first stage of sleep, the amplitude is greatly reduced. The dominant rhythms are very mixed but still alpha-like, around 12 to 14 Hz. As the stages progress, the dominant frequency decreases further to around 2 Hz (delta waves), while the amplitude increases quite clearly into the high-amplitude peak or spindles, typical of deep sleep, which this person reaches about fifty minutes after going to bed. Though asleep, the person still moves around and shifts in position, so the muscle tone is still active. Until now there is no dreaming. All of this time the subject is in non-REM sleep.

"The next thing that happens in the human sleep pattern is that the stages reverse, progressing back from stage four to stage three, then stage two. Then one enters into a completely different state, the REM or paradoxical sleep that involves dreaming. In the first two to three hours of the night, the pattern of moving in and out of deep sleep predominates. But as dawn approaches, REM periods tend to predominate and deep sleep disappears. Thus sleep is not just a single state, nor are its variations random. It is a pattern that is highly regulated over time and that includes distinct states of human consciousness."

It was time for some clarifications. His Holiness asked: "Does this happen for everyone? What is the time factor in shifting from stage one to two to three and so forth?"

"Yes, this is a very basic human mechanism. And the time for transition is very quick. You can go from awake to the first stage of sleep in five minutes, for some people fifteen or twenty minutes. Sometimes the transition from stage one to REM can be much quicker, but you always pass through the four stages and up again. There's more variability from awake to stage one than there is returning from stage four back up."

"It is clear that there are variations in going from a waking stage to stage one, but when you're at stage one going to two, three, four, is there a general rule for everybody or is there also variation from one person to another?"

"The transition times are variable. What is not variable is that you cannot skip a stage."

Characterizing REM Sleep

I continued with the presentation. "Let me explain more fully what is meant by REM. In fig. 2.4 we see the electrical signals from two people (A and B) recorded just as they slip from awake to stage one. In addition to the sample EEG, two other electrical signals are picked up externally: the EOG is the so-called electro-oculogram that shows eye movement; the EMG is the electro-myogram that indicates the level of motor control in skeletal muscle.

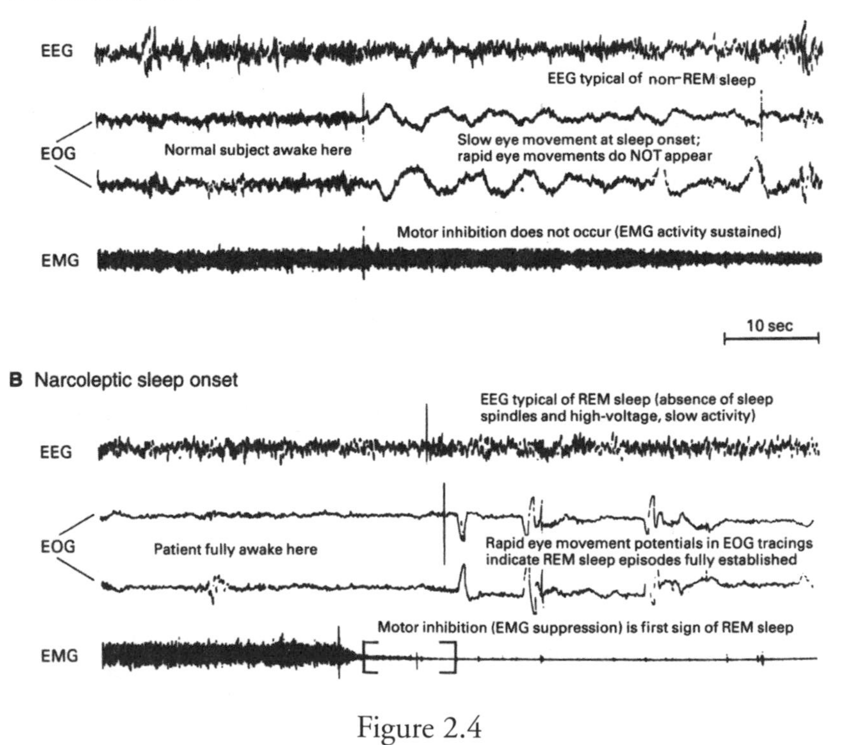

Figure 2.4

Electrical signals recorded at the onset of sleep for a normal subject and a narcoleptic subject. (From Kandel, Schwartz, and Jessel, Principles of Neural Science, *3rd ed., Norwalk, CT: Appleton & Lange, 1991, p. 814. Printed with permission from Appleton & Lange.)*

"The patterns shown by A in figure 2.4 are those of a normal individual. Notice that the EEG does not change very much at the transition point. It's just a little smaller in amplitude. In contrast, the EOG changes radically, slowing down to indicate the slow eye movements of rolling one's eyeballs. The muscle tone changes very little.

"In B we see the patterns of a person who suffers from narcolepsy. Narcoleptic people cannot control the urge to go to sleep, even in the middle of a conversation or dinner. It's a very embarrassing, very disruptive condition. In this case, right after the onset of sleep the EOG shows the very rapid twitches known as rapid eye movement. The muscles lose their tone and go limp, hence there is a flat EMG. The EEG, on the other hand, remains virtually unchanged, which is why early researchers called this 'paradoxical sleep': the EEG seems awake but the person is asleep. These three characteristics of awake-like EEG, REMs, and flat EMG are typical of what is called REM sleep. This state is not normally entered unless you go from slow sleep stage four back up, as we saw. But narcoleptics enter REM directly, which is not normal. However, they do provide an excellent demonstration of all conditions of sleep in one continuous recording!

	Non-REM	REM
EOG	slow rolling eye movements	rapid eye movements
EMG	moderate activity	atonia in peripheral muscles
cerebral activity	decreased	increased
heart	slows down	no change
blood pressure	decreased	no change
brain flow	no change	increased
respiration	decreased	increased, variable

Figure 2.5

A summary of the differences between non-REM and REM sleep.

"For a more complete picture of these various states of human consciousness, let us compare non-REM and REM sleep in contrast to the awake state (fig. 2.5). By examining characteristics such as cerebral activity, heart rate, blood pressure, brain blood flow, and respiration, you can see that these two types of sleep are radically different configurations of the entire body. Cerebral activity is a measurement of the global amount of brain electrical activity. In non-REM sleep, the brain becomes more silent. Interestingly, in REM sleep it's more active than in wakefulness. This goes completely against the old intuition that sleep is passive. Relative to wakefulness, the heartbeat in non-REM sleep is a little slowed down, and unchanged during REM. Brain blood flow is the total amount of circulation of blood in the brain, which is a measurement of how much oxygen and nutrients are needed. This clearly increases in REM sleep, again an indication that this is a very active process. Finally, respiration in non-REM sleep slows down a little bit, while in REM sleep it's very variable. In summary, there is a very clear and distinct pattern here. Within non-REM sleep there are stages that progress in a continuum, but there is a radical shift from non-REM to REM sleep."

Dreaming and REM

"Why is the REM sleep state so important? The brain patterns of REM sleep correspond to the dream state. When you wake people up from REM sleep, more than eighty percent say that they were dreaming, and they can tell you what they were dreaming about. If you wake them up from stage four, less than half do."

"Do you mean that even during the non-REM stage the person can be dreaming?" His Holiness asked.

I was expecting that question. "Yes. It depends on how you evaluate the subjective report, but it is generally accepted that about half of the people who are woken up from non-REM sleep report dreaming or mental activity. Many say they were thinking rather than dreaming. They report some kind of experience or mental activity, but it doesn't usually have the same full-fledged story quality as a dream."

"Is there only a tenuous relationship between the REM and the dream state?" he insisted.

I could try to qualify the answer in a fuzzy way. "It depends on how you set up your criteria. From stage four people may say, 'I was thinking about something' or 'I was considering something' but from non-REM sleep, people very rarely report a complete vivid story such as 'I was flying like an eagle and I saw my home.' In non-REM sleep, it's more like mental content than like a movie. Even in going from awake to sleep, one sometimes has very brief flashes of imagery, called hypnagogic dreaming. Such sudden bursts of imagination, whether visual or auditory, also occur in people left in darkness. So it's not fair to say that dreaming occurs only in REM sleep because other kinds of dreamlike experiences occur in all of the other stages. But it is clearly true that vivid, visual, storylike dreams occur classically in REM sleep.

"We spend about twenty to twenty-five percent of a complete circadian cycle in REM sleep. From the neuroscientific point of view, therefore, we dream every night, although we often are not dreaming. The classical sequence of sleep patterns usually happens a second time during the day. This is called the biphasic sleep cycle. At nine A.M., after a good night's sleep, a normal young adult takes about fifteen minutes to go back to sleep. But everybody knows that at two o'clock in the afternoon, at siesta time, it's easy to fall asleep. Also, the time that it takes to fall asleep is typically shorter by about five minutes in an older person."

Sleep in Evolutionary Perspective

"I want to give two arguments for the tremendous importance of REM sleep in the history of animal life. If REM sleep or dreaming were strictly human, then animals wouldn't do it. But it is fascinating that other primates have almost the same pattern of sleep as humans. They have the same kind of cycles and they go through the same stages. In our next nearest relatives, the situation is even more interesting, because virtually all the large mammals have REM and non-REM sleep."

The Dalai Lama immediately jumped in to ask what the exception was. I said that the anteater doesn't have REM sleep. "Maybe it's their diet!" he laughed.

"It is also remarkable that humans typically sleep lying down. Cats sleep curled up. Most dogs sleep sprawled out. The tiger likes to put himself on a tree. The elephant sleeps standing up, the hippopotamus under water. Cows can sleep with their eyes open. Dolphins continue to swim, as only half of their brain sleeps! Some animals have short periods of sleep. For example, the elephant sleeps on the average only 3.2 hours of the day. Rats sleep 18 to 20 hours of the day. There is an interesting relationship: the smaller you are, the more you sleep.

"Some animals like rats pass very quickly from awake, through the four stages, and then pop into REM very quickly, and they have very short periods of REM sleep. Some animals have very long REM sleep periods. There is quite a bit of variability. Some animals, such as dolphins or cows who sleep standing up, do not lose muscle tone in REM sleep. So although the existence of REM and non-REM sleep is universal among mammals, its expression adapts to their particular lives. To a biologist, this means that evolution has made an incredible effort to restructure the brain many times to keep REM and non-REM sleep, and to shape it in different modes. Walking, standing up, changes in posture: the pattern varies, but still we find REM and non-REM sleep. The appearance of the same basic state in many multiple forms indicates that there is something very important here, because evolution did not let any mammal, except for the anteater, lose it.

"What about other animals, beyond the mammals? How far can we go in evolution and still find REM and non-REM sleep patterns? Both birds and mammals evolved from reptiles. Birds, which sleep mostly standing up, do have REM sleep. Some biologists suspect that birds that migrate for many days sleep while flying, just as the dolphin sleeps while swimming. Dreaming while flying over the planet."

His Holiness opened his eyes wide and asked, "Is that verified?"

"No, this is a hypothesis. Since they have cycles of REM, and some fly for days, it's a logical inference. Birds apparently have reinvented REM sleep independently, because reptiles don't seem to display it. Although there is no evidence of REM in sleeping reptiles, things get a little complicated here. We use electrodes in the cortex to recognize the typical brain waves of REM sleep, but reptiles, like all other premammalian groups, don't have the same kind of cortex as we do. The same kind of cells are not aligned in the same manner, so it's not entirely clear whether REM sleep takes place in reptiles. But from the level of reptiles on up there is no question that everybody sleeps and dreams in patterns of REM, non-REM, and awake. For a biologist there is an impressive argument here for the fundamental nature of sleep and dreaming."

Why Do We Sleep?

The Dalai Lama moved on to the logical next step: "Has it been identified exactly what REM sleep, that has proved so vital in the course of evolution, does for you physiologically?"

"That's the key question: Why do we sleep and dream? What is the purpose of it? There is quite a debate in neuroscience about this, but there are fundamentally two ways of answering the question. Some people think of sleep as a form of restoration or replenishment. But while this feels intuitively true, nobody so far has identified precisely what it is that we are replenishing. You spend a lot of energy during sleep; there is actually more oxygen consumed during REM sleep than when awake, so it's not a simple matter of letting the machine cool off. Because REM is such an active state, it's not obvious how we are replenishing, restoring, or refreshing ourselves.

"The other answer, which I personally prefer, is that REM sleep is a fundamental cognitive activity. It is the place where people can engage in imaginary play, trying out different scenarios, learning new possibilities; a space of innovation where new patterns and associations can arise, where whatever was experienced can be reelaborated. This is quite close to some views in psychoanalysis. Dreaming provides a space where you don't just cope with immediacy, but instead can reimagine, reconceive, reconceptualize. It's a form of rehearsal that allows you to come up with new possibilities. I'd love to know whether this is considered the nature of dreaming in Buddhism as well. For animals like reptiles and insects that don't learn very quickly or change behavior much, this is probably not too important, although we should be very careful here. We cannot ascertain whether insects sleep because they don't have a cortex.

"Another line of evidence that seems to support the cognitive interpretation is the pattern of dreaming over a lifetime (fig. 2.6). Premature babies sleep in REM up to eighty percent of the time, and newborn infants spend fifty to sixty percent of the time in this state. As we all know, babies sleep fifteen to twenty hours of the day. REM sleep seems to be necessary while growing up, physiologically and mentally. To me, this is an argument for the importance of cognitive imagery in dreaming. Beyond sixty-five years of age, one sleeps and dreams much less.

"This is a very tempting idea, but it is by no means the consensual, standard answer. Some people have more extreme notions: that dreams are just random neuronal firing with no meaning whatsoever, or that dreaming and sleeping have to do with energy conservation by avoiding movement. It's not a simple issue, because an animal that is alert and awake also conserves energy. The question is still much in debate."

I needed to bring the presentation to an end. "Sleep research is a large and active field, and new discoveries are further elaborating the phenomenology we have been discussing.[4] For example, recent results clearly show that different groups of neurons in the brain turn on REM sleep, non-REM sleep, or wakefulness. The neurons are mostly in the brain stem and the cortex, where commands to change the activity states of muscles or eyes originate. We can artificially manipulate which groups are active in animals. The onset of REM, non-REM, or wakefulness each corresponds to a different kind of transmitter substance, but they are complex patterns, not just simple switches.

"With such a complex brain system many things can go wrong. There are three main types of sleep disorders: insomnia, where people tend to sleep too little and do not enter non-REM easily; hyper-

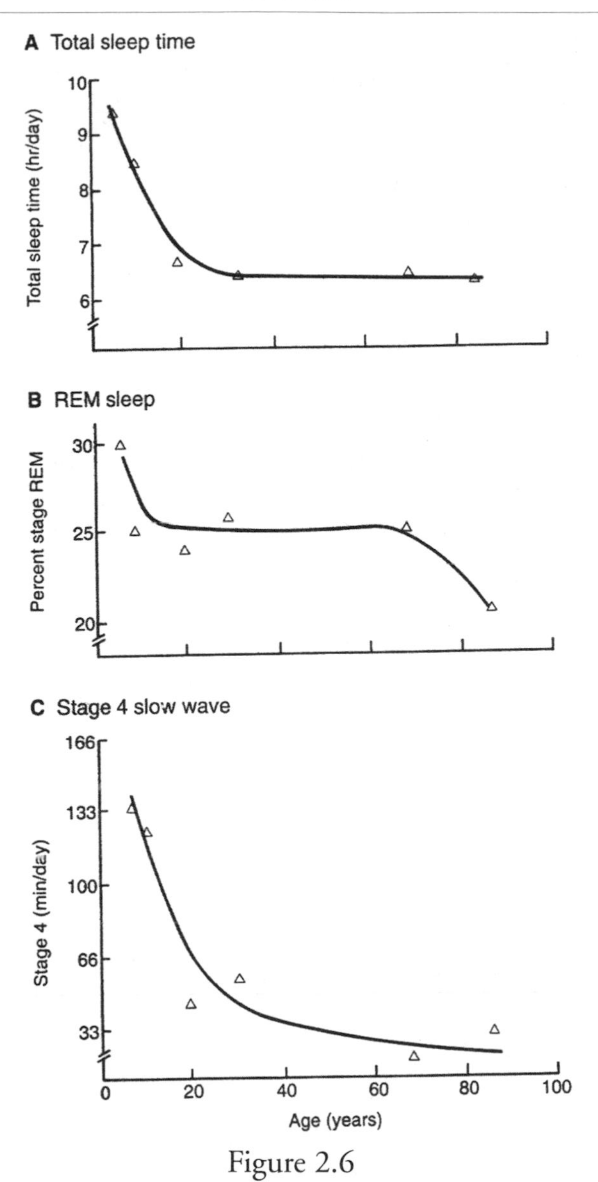

Figure 2.6

The amounts of time spent in sleep generally, and in REM sleep, decrease with a characteristic pattern from birth to old age. (From Kandel, Schwartz, and Jessel, Principles of Neural Science, *3rd ed., Norwalk, CT: Appleton & Lange, 1991, p. 796. Printed with permission from Appleton & Lange.)*

somnia, where people sleep too much as in narcolepsy; and parasomnia, which is neither sleeping too much nor too little, but in disrupted patterns. An example is sleepwalking, or somnambulism.

Some of these complex and varied sleep disorders are physiological, whereas others are psychological. When one is stressed or depressed, sleep is affected. Similarly, if one has trouble sleeping, one gets slightly mentally disarranged."

Dreams in the Tibetan Tradition

I had finished. We all stopped for a moment to gather notes and sip some tea. I then launched into the discussion stretch of our session. For me this was an especially difficult moment because science thrives on an impersonal voice that is intrinsic to the scientific method. In this light, first-person accounts and individual insight do not have scientific "objectivity" in the classical sense. This is why some Western scientists feel awkward when they approach the notion of consciousness, which is eminently and irreducibly first-person. One aspiration of this Mind and Life meeting was to explore in a nondogmatic way how such Western reactions can eventually be overcome with new ideas or new methods that respect both scientific and experiential observations.

For ten centuries, the Tibetans have been involved in the phenomenology of dreaming. One of their particularly important traditions originated with the eleventh-century Indian yogi Nāropa and was later transmitted to Tibet under the name of the Six Yogas of Nāropa. One of these yogas deals elaborately with dreams and dreaming, and later practitioners and theoreticians improved on this knowledge base so that it became a refined art.

"Having given His Holiness this brief account of sleep physiology, I am curious about the meaning of sleep and dreams in the Tibetan tradition. Is there an idea of different levels of consciousness being the source of different kinds of dreaming? Is there an answer for why we dream?"

His Holiness answered: "There is said to be a relationship between dreaming, on the one hand, and the gross and subtle levels of the body on the other. But it's also said there is such a thing as a 'special dream state.' In that state, the 'special dream body' is created from the mind and from vital energy (known in Sanskrit as *prāṇa*) within the body. This special dream body is able to disassociate entirely from the gross physical body and travel elsewhere.

"One way of developing this special dream body is first of all to recognize the dream as a dream when it occurs. Then, you find that the dream is malleable, and you make efforts to gain control over it. Gradually you become very skilled in this, increasing your ability to control the contents of the dream so that it accords to your own desires. Eventually it is possible to dissociate your dream body from your gross physical body. In contrast, in the normal dream state, dreaming occurs within the body. But as a result of specific training, the dream body can go elsewhere. This first technique is accomplished entirely by the power of desire, or aspiration.

"There is another technique that arrives at the same end by means of *prāṇa* yoga. These are meditative practices that utilize the subtle, vital energies in the body. For these techniques also it is necessary to recognize the sleep state as it occurs.

"It seems that some people have this ability naturally, without any specific practice. For example, last year I met a Tibetan living in Nepal who told me a story about his mother. Some time ago, his mother told those around her that she was going to be immobile for some time, and not to touch or disturb her body. They didn't mention whether she was breathing or not, but for one whole week her body was totally immobile. When she woke up, she said that she had visited various places while her body was immobile. In other words, she had an out-of-body experience with her dream body. So in the special dream state, it seems that one is using a very subtle body which disengages from the gross body and is able to travel independently."

The answer seemed to move too quickly to the fringe of experience, touching on unusual concepts such as "out-of-body experience" and the "dream body." Most of us in the West have not been exposed to experiences of disembodied dream bodies, and I feared that we might find ourselves lost in cross-cultural non-talk. One of the main purposes of our gatherings was to stake out a common ground on which both traditions would be able to stand independently. This is one of the most important challenges of this whole series of meetings, and the last interaction encapsulated the challenge perfectly. Thus I instinctively moved the discussion back to a potential common ground: "Is there a distinction between recognizing the dream as it occurs in REM sleep, and recognizing sleep in a non-REM state?"

His Holiness's answer referred to the body of advanced Tibetan Buddhist teachings known as the Vajrayāna, or Diamond Vehicle for human awakening. "You spoke earlier of the four stages in non-REM sleep that precede the REM sleep," he said. "In Tantric Buddhism or Vajrayāna, there are four stages in the process of falling asleep, culminating in the so-called clear light of sleep. From that clear light of sleep, you arise into the dream state of REM sleep." An amused expression appeared on his face, followed by a contagious broad smile that brought laughter from all of us. "You said that the four sleep stages occur in a definite, unalterable order. A person untrained in meditation can't tell whether the four stages described in Vajrayāna Buddhism are unalterable. However, a person who is well trained in Vajrayāna meditation can recognize a strict order in these four states of falling asleep, and is well prepared to ascertain an analogous order in the dying process. It's easier to recognize the dream as the dream than to recognize dreamless sleep as dreamless sleep. If you can recognize the dream state while in it, then you can visualize and deliberately reduce the grosser level of mind to return again to clear light sleep. At that point the subtlest level of mind—the clear light of sleep—is easier to ascertain."

"But while falling asleep," I interjected, "normal human beings simply black out, and no ascertainment of any kind is possible."

"It's true," he replied. "Going though this transition without blacking out is one of the highest accomplishments for a yogi. But there may be one difference between sleep physiology and the Tibetan tradition. According to Vajrayāna Buddhism, these four stages are said to repeat in reverse order when you awaken from the dream state. It happens quite quickly. You have not mentioned anything like that in neuroscientific sleep research. You did mention that from the REM state you go to an awake state, but what about repeating the cycle of various stages in between?"

That was an interesting observation on the data I had shown. Once more, I put up the graph showing the sequence of stages (fig. 2.3). "One goes straight from REM to awake. REM is the closest to waking that we know from experience. During the sleep cycle, you don't go into deep, slow sleep in the later part of the night. You stay in shallow sleep and simply go from REM sleep down to stage two, then up to REM and back to stage two again. According to the scientific view, if you are in stage four, you must have gone through stage three and two. If you are in stage three you must have gone through stage two and one, but you can easily go from REM to stage one and from stage one to REM."

"Perhaps the view presented in sleep physiology is closer to an earlier view held by Tibetan scholars of an older school," said the Dalai Lama, reminding us that Buddhism, like most of today's active traditions, has evolved greatly from its founding sources. "The four stages culminate in the clear light, and then you have the first three stages in reverse, coming back from the clear light to the dream state. In the later writings this view seems to change, especially in relation to the intermediate *bardo* state and the tradition of the Six Yogas of Nāropa, whose sources go back to Marpa."

Nāropa, a famous tantric adept who lived in India in the eleventh century, was the teacher of Marpa, a Tibetan who traveled twice to India to receive teachings and bring them back to his native country. Marpa later became the main founder of the New Translation lineage of Tibetan Buddhism, which has since evolved into many different schools, some of which are still active today. Such differences within Buddhism remind me of the scientific world, where differing views also coexist for long periods of time without resolution.

"Incidentally," said His Holiness, "Tibetan Buddhism considers sleep to be a form of nourishment, like food, that restores and refreshes the body. Another type of nourishment is *samādhi*, or meditative concentration. If one becomes advanced enough in the practice of meditative concentration, then this itself sustains or nourishes the body. Although sleep is a source of sustenance for the body, it's not clear how dreaming serves the individual, other than its use in meditative practice. In Buddhism, the origin of dreams is understood as an interface between different degrees of subtlety of bodies—the gross level, the subtle level, and the very subtle level. But if you ask why we dream, what's the benefit, there's no answer in Buddhism."

He then turned to a question that had raised his curiosity: "We have a distinction between REM sleep and the fourth stage of non-REM sleep. While you're in REM sleep, does the state of consciousness of the fourth stage of non-REM continue simultaneously, or is that interrupted?"

"These are different states," I answered. "When I am in REM, I have one mode of consciousness. When my mind-brain changes, I have another one. I don't need to posit that the other one continues. I prefer to think these are emergent properties of the brain and body configuration."

"Must REM sleep necessarily be preceded by non-REM sleep?" he insisted. "According to Tibetan Buddhism, to be in the sleeping state presupposes that the mental factor of sleep has manifested, and sleep can occur with or without dreaming. But if dreaming occurs, the mental factor of sleep must be present. The mental factor of sleep is the basis for dreaming as well as dreamless sleep. In one text, a Tibetan scholar makes the almost contradictory statement that in deep sleep there is no sleep, because there is no awareness or consciousness. Thus, sleep, as one of the mental factors, is not present in deep sleep."

He looked at me for an answer. "Neuroscientists would say that perhaps you would have to define two mental factors, REM sleep and non-REM sleep. But if we call spontaneous mental or visual images dreaming, then dreaming occurs in all three states: wakefulness, REM, and non-REM. You can be awake and hallucinate; you can have hypnagogic images when you're falling asleep; you can have dreams with mental content in non-REM sleep; and you can have classic dreams in REM sleep. But if we define dreaming strictly as vivid, storylike, with a continuous plot, then it's more of a REM phenomenon." His Holiness nodded and seemed to reflect on this suggestion as to how science and Abhidharma, the Buddhist theory of mental functioning, can modify each other.

Dissolution in Sleep and Death

Pete Engel, eager to discuss sleep, dreams, and death in more general terms, referred to the book *Death, Intermediate State, and Rebirth in Tibetan Buddhism* by Lati Rinbochay and Jeffrey Hopkins (London: Rider, 1979). In preparation for the meeting, I had circulated a large volume of readings, some published by the meeting's participants, some from Tibetan sources, and still others on our topics of interest. Lati Rinbochay and Hopkins's book explores the controversial idea of after-death experiences in the intermediate state following death and preceding rebirth, known as *bardo* in Tibetan.

Pete began in his low-key tone, "The book discusses stages of death which are then repeated in reverse order in the *bardo* state during the transition towards rebirth. It says that sleep is in essence a rehearsal for this process of dying, and I was struck by the similarities between sleep and death. It also states two other conditions in which those same steps take place: the meditative state and orgasm. I'd like to know more about this, because the meditative state is quite different from the sleep state neurologically, and I'm lost when you throw in orgasm! I am struck by the fact that there are comparable steps in the Buddhist concept of sleep and in the scientific observations of sleep, but what are the steps in the meditative state that may be similar to sleep in the Buddhist view, and how does orgasm fit into this?"

"The experiences that you have while falling asleep and while dying result from the dissolution of the various elements," answered the Dalai Lama. "There are different ways in which this process of dissolution takes place. For instance, it can also occur as a result of specific forms of meditation that employ the imagination. The dissolution, or withdrawal, of the elements corresponds to levels of subtleties of consciousness. Whenever this dissolution occurs, there is one common element: the differences in the subtlety of consciousness occur due to changes in the vital energies."

He explained the methods behind these concepts. "There are three ways that these changes in the vital energies can occur. One is a purely natural, physiological process, due to the dissolution of the different elements, namely earth (solidity), water (fluidity), fire (heat), and air (motility). It happens naturally in sleep and in the dying process, and it's not intentional. An analogous change occurs in the vital energies as a result of meditation that uses the power of concentration and imagination. This change in the vital energies results in a shift of consciousness from gross to subtle. The third way is through sexual intercourse. However, the shift of energies, and the shift from gross to subtle consciousness, does not occur in ordinary copulation, but only through a special practice where one controls the movement of the regenerative fluid in sexual intercourse, both for men and women."

Pete persisted, "Is the end result—the dissolution of earth, water, fire, and air—different or the same in these practices?"

"It's not exactly the same," the Dalai Lama said. "There are many different levels of subtlety in the clear light experience. For example, the clear light of sleep is not as deep as the clear light of death. Vajrayāna Buddhism speaks of five primary and five secondary types of vital energy, as well as gross and subtle aspects of those two sets of five. In the clear light of sleep, the grosser forms of these various energies dissolve, or withdraw, but the subtle forms do not. As an indication of this, the person continues breathing through the nostrils."

Since we were going to explore death in the second part of the week, I was concerned that the discussion was moving too far afield ahead of time, and asked Pete to keep the focus on sleep and dreaming. He agreed: "I'm more interested now in the similarities between sleep and meditation. If it is possible through practice, though very difficult, to go through the stages of sleep and reach the clear light of sleep intentionally, how is that different from meditation?"

His Holiness answered, "It is important to realize that there are many forms of meditation. These issues are not even discussed in the lower three classes of Buddhist tantra, only within Highest Yoga Tantra. Dream yoga is a discipline all unto itself."

I was curious. "Can one really engage in it without a whole foundation that precedes it?"

"Yes, it is possible without a great deal of preparation. Dream yoga could be practiced by non-Buddhists as well as Buddhists. If a Buddhist practices dream yoga, he or she brings a special motivation and purpose to it. In the Buddhist context the practice is aimed at the realization of emptiness. But the same practice could be done by non-Buddhists."

Are There Correlates of Subtle Mind?

I asked a question that had been at the back of my mind, and I probably spoke for many of us. "Suppose somebody has practiced to the extent that he can consciously go through the stages of sleep and remain in the clear light on a regular basis. If we did the same experiments on him that I described, would you expect any external changes? Would any of the signs by which we recognize REM and non-REM sleep be different? Would stage four appear changed?"

In his answer to this and other questions, His Holiness repeatedly used the term *prāṇa.* As our interpreter Alan Wallace pointed out, it is better translated as "vital energy" rather than "subtle energy." The former term may mistakenly evoke the vis vita and elan vital of medieval and Renaissance Europe, but it still seems more accurate than calling *prāṇa* merely "subtle." Moreover, there are three levels of *prāṇa*—gross, subtle, and very subtle—so this further confuses matters. Finally, *prāṇa* is by nature confined to living organisms, so "vital" seems thoroughly appropriate.

"It is difficult to say whether one can obtain external correlates of the state of clear light. We would expect very little disturbance of the body's vital energies while a person is abiding in the clear light of sleep in the fourth non-REM state. I think the term clear light of sleep derives from meditative experience. In sleep yoga and other practices that use very subtle states of consciousness to gain insight into emptiness, there are experiences of clarity and luminosity."

Charles Taylor interrupted, "So this training culminates in the ability to remain in the *bardo* and not be carried away by the different stages? Or did I get it all wrong?"

His Holiness smiled, as usual. "You have to relate this to a broader picture of the Buddhist path. We speak of different embodiments of a buddha, including the Sambhogakāya, the very subtle body of an awakened being, and the Dharmakāya, the enlightened mind of an awakened being. The practice of developing the special dream body is ultimately aimed at achieving the Sambhogakāya, whereas the ultimate purpose of ascertaining the clear light of death is achieving the Dharmakāya. The Sambhogakāya is an illusory body, or physical form in which a buddha appears to others, while the Dharmakāya is self-referential, directly accessible only to a buddha. So the practice of dream yoga relates to the Sambhogakāya, and the practice of the clear light of sleep relates to the Dharmakāya."

Alan stepped out of his role as translator and asked, "Does REM sleep differ for a person who is familiar with language, and someone, an adult even, who is not? Does the mind operate differently if language-based concepts don't arise?"

"It would be very difficult to test. If someone is not acquainted with language you cannot ask them," I said, and we all laughed. "This is where we run into the limitations of current method. But babies and other mammals do have similar patterns."

His Holiness continued, "One of the five primary types of vital energy is called pervasive energy. I wonder if the power of this pervasive energy throughout the body possibly increases during REM sleep and decreases during non-REM sleep. It's quite possible that this would have a relation to conceptualization."

I reflected, "Pervasive energy is not a concept familiar to science, but the heart flow and cerebral blood flow that increase during

REM might be indices of pervasive energy."

Next His Holiness suggested an interesting experiment. "Have you ever conducted EEG research on a dying person who is asleep? For example, do you know how long REM sleep lasts while a person is in the process of dying?"

"One criteria for declaring somebody dead is that the EEG altogether begins to flatten out and disappear," I said. "There are no more oscillations, so there is no way to tell REM from non-REM."

"When the brain activity is finished," he insisted, "it's an open question whether there is still pervasive energy or not. There seem to be three criteria for death: heartbeat, respiration, and brain activity. How many minutes does the brain function after the heartbeat has stopped?" We all agreed that it was a mere few minutes. "Can REM occur during those few minutes?"

Pete answered, "I don't think that that's ever been observed or tested. At that point, the eyes become fixed or roll up, the pupils become dilated, and there is no brain activity. The time it takes to die after the heart stops beating can be extended by cooling the body. People who drown in very cold water may still be revived after fifteen to twenty minutes, even though the EEG is flat and they do not breathe or have any heartbeat. Drugs can also reproduce this flat EEG when the patient is not dead."

We were clearly reaching the edge of what could be envisaged in this regard. But the non-answers were very interesting. I added, "The moral to remember here is that measurements of EEG are very, very gross. There is no contradiction in assuming that there is much more linguistic, associative, and semantic activity going on in humans during REM sleep. In animals that don't have REM sleep, there may be a different kind of cognitive activity, but you won't see that in the EEG, which is just too gross a measure. When somebody is dying and the EEG flattens out, that doesn't mean that there is nothing going on. There might be lots of things going on. The same applies to the earlier question about the difference between a normal individual at stage four and somebody who can remain conscious in the clear light of sleep. Maybe EEG measurements won't show a difference, but more refined methods in the future could."

Intention and Effort in Practice

At that point Jayne Gackenbach steered the conversation to a topic that was directly related to her work. "Is the intention in ascertaining the dream and then controlling it that at some point you let go of it?"

"I'm not sure whether you let go of control," replied His Holiness. "Clearly, in order to maintain the practice of dream yoga, you need a certain degree of effort and intention, and you must sustain that intention. As you become more familiar with the practice, less and less effort will be involved, because you're getting very good at it. But there are phases of both Buddhist and non-Buddhist meditative practice during which effort is utterly suspended. One example of that is in the practice of Dzogchen, or the Great Perfection. This entails a very special kind of suspension of effort. There's another practice common to Buddhists and non-Buddhists, in which you simply abide in equanimity. But that is different from the suspension of effort in the Buddhist practice of Dzogchen."

"Is that abiding in equanimity the objective of the effort? Or does controlling the dream eventually lead to serenity and the dream is just there?" persisted Jayne.

"This practice of just releasing effort in equanimity is a separate practice unto itself. We're talking about two different forms of practice here. In the first you need to recognize the dream state and then control the dream. You do that for a specific purpose, which is to generate the subtle dream body, which can be separated from the gross body. In the second practice, you cultivate the ability to experience the clear light of sleep, and for this, controlling is not necessary. The two practices are really quite distinct and are used for different purposes. The purpose of the practice of applying effort to recognize the dream, and intentionally transforming and controlling its contents, is to acquire the special dream body. That dream body can then be used for a wide variety of purposes. This practice is analogous to recognizing the intermediate state (*bardo*) as the intermediate state. The major challenge that faces you in this practice is to sustain your recognition of the intermediate state and not be overwhelmed when various apparitions appear to you. That's a challenge. On the other hand, the practices that lead to the realization of the clear light of sleep are a preparation for recognizing the clear light of death."

Sleep, Orgasm, and Death

Joyce McDougall added an interesting remark from her own profession. "Psychoanalysis may offer a comment on the relation between sleep and orgasm, which can both be linked imaginatively to the idea of dying. People who suffer from insomnia and people who cannot achieve orgasm may discover in the course of analysis that their inability to fall asleep or to fuse with someone they love in an erotic union derives from a terror of losing the sense of self. It's interesting, too, that in France, orgasm is called *la petite mort*, the little death. In Greek mythology, sleep and death are brothers, Morpheus and Thanatos. To let yourself sink into sleep you have to let go of your personal idea of self and dissolve into the primal fusion with the world, or with the mother or the womb. Losing the everyday self is experienced as a loss instead of an enrichment. This can also apply to people who cannot enjoy orgasm. It relates to what Professor Taylor was talking about. The willingness to lose our sense of self that allows us to sink into sleep or orgasmic fusion can also allow us to be unafraid of dying. We might say that sleeping and orgasm are sublimated forms of dying."

Laughing, His Holiness said that in Tibet the best solution for those who have such fears was to take ordination. On a more serious note he added, "In the Tibetan Buddhist literature, it is said that one experiences a glimpse of clear light on various occasions, including sneezing, fainting, dying, sexual intercourse, and sleep. Normally, our sense of self, or ego, is quite strong and we tend to relate to the world with that subjectivity. But on these particular occasions, this strong sense of self is slightly relaxed."

Joyce elaborated, "Is there a link between the difficulty of letting go of the self in the wide-awake world, and being unwilling or unable to let go of the gross body image and allow the more spiritual body image to come? Would clinging to the gross body image stop this letting go?"

His Holiness replied, "I think there might be some correspondence there, because our sense of self is very much related to our bodily existence. Actually there are two senses of the self, one gross and one subtle. The gross sense of self arises in dependence upon this gross physical body. But when one experiences the subtle sense of self, the gross body becomes irrelevant, and fear of losing one's self vanishes."

Awareness and Discontinuities

At this point Joan Halifax observed that the existence of different states implies transition between them: "All of these states seem to involve a momentary cessation or eclipse in awareness, so that something actually does seem to die. Whether it's the gross or subtle level of the body, there is a break in continuity, a gap after which the continuity reconstitutes itself. Is it not true that one aspect of the practice is to maintain the continuity beyond the gross or even the subtle level—at a level which has no conditions? To maintain the continuity of something that is nothing." We all laughed with her struggle to phrase the point. "We don't have the words for it in our language."

His Holiness answered, "There are states of meditation in which you simply have a sense of emptiness, and at that time you don't have even a subtle sense of self. Although you have no sense of 'I' at that time, this doesn't mean that there is no 'I' then. The Tibetan term for consciousness is *shes pa*, which literally means 'knowledge' or 'awareness.' The etymology implies awareness of something, and this defines consciousness at the gross level. However, at subtler levels there may not be an object of awareness. This is analogous to the paradoxical state of 'thoughtless thought.' This is a conceptual state of awareness free of certain types of thought or certain levels of thinking; but 'thoughtless' here does not mean totally devoid of thought."

"Do you mean the absence of cognizance?" I ventured. His Holiness paused and asked in turn, "Can you, Francisco, distinguish between thought, awareness, and cognizance?"

"*Cognizance* refers to a quality of ascertainment or discernment. You can have that quality without necessarily having a *thought*, which always has a particular semantic content. Awareness has multiple meanings; one is cognizance, but it could also range into more subtle levels. Both *awareness* and *cognizance*, unlike *thought* and *thinking*, seem to be extendible in use to more subtle levels of consciousness such as nonintentional thought, or thought without an object. In cognitive science people shy away from the use of *awareness* and prefer to talk about *cognizance* and *cognition* as long as there is a content to the knowing."

Charles had a stricter interpretation: "In my understanding, *cognizance* and *knowledge* have the sense that there is something, some content, that you know or grasp. In contrast, I can be *aware* of something without knowledge of it. *Cognizance* is an achievement. That's why I find it hard to understand the idea of objectless cognizance."

"It is a state that has no content and no propositional object," advanced His Holiness.

Charles was not daunted. "It's hard to think of it as not having an object, though I can understand how there could be a paradoxical state that we would be forced to call 'awareness of nothing.' We may be making a mistake in trying to look beyond ordinary human states, because the words were designed for ordinary human states where there is no contentless awareness and no contentless cognizance." I could only point out that that should not stop us from postulating that such human capacities could exist. Charles conceded, "In all traditions we have to twist, and pull, and push ordinary language to capture states that are not ordinary."

His Holiness assented, "This is true also in the case of Buddhism. It's just the nature of language."

The day had breezed by. It was five o'clock. His Holiness thanked us and left with a bow. It was time to go back to our cottage and continue informally to trace the many threads that had been uncovered.

3

Dreams and the Unconscious

Psychoanalysis in Western Culture

WHETHER ONE IS A DETRACTOR or admirer of the psychoanalytic tradition, one thing is certain: Freud and his followers have radically transformed our Western understanding of what a mind is, of what it is to be a person, and of therapeutic intervention. A multitude of psychological theories and clinical approaches now exist, some quite sophisticated, others more superficial. In North America the flair and the search for variety have been far greater than in Europe or South America, where psychological theory and practice remain, to a great extent, psychoanalytic.

As organizer it was my responsibility to make sure that the Dalai Lama and the Tibetans sampled a fair representation of Western perspectives regarding the shadow zones of the ego. Psychoanalysis has introduced into common knowledge at least two key ideas for this meeting: the notion of the human unconscious and its depths, and the central role that dreams play in exploring the human psyche. Admittedly, psychoanalysis is not part of mainstream science, nor does it pretend to be. Yet it was born out of neurology and psychiatry, and plays an important role as the basis of many treatment centers throughout the Western world. Furthermore, the advent of the cognitive sciences has done much to renew the links between psychoanalytic pragmatics and theory and science.[5]

When searching for somebody who could represent this tradition with experience and authority, the name of Joyce McDougall came naturally to mind. Born in New Zealand where she obtained a doctorate in education, she went to London to undertake psychoanalytic training, and studied for some years at the Hampstead Clinic for Child Psychotherapy, where Anna Freud was the leading inspiration. Since her husband's work took him to France, she continued her training for adult psychoanalysis in Paris, where she has been at the center of the theory and teaching of this discipline for twenty-five years. Her books are an example of lucidity and open-mindedness, qualities not always found in psychoanalysis. In a recent work, she weaves together many strands of analytic practice in what she calls "theaters" of the body and the mind.[6]

For the second day of our meeting, on a beautiful sunny morning, Joyce took the presenter's seat. It was the first of the two occasions in this meeting when we moved from hard brain science to a field in which human experience and its drama were central.

Freud and Company

Joyce opened by saying, "It is my honor and pleasure to try to communicate something of the science and the art of psychoanalysis." The juxtaposition of art and science was certainly what was needed. "Sigmund Freud was the founder of this science and of its therapeutic art. At the end of the last century, a century of conservatism, of dominating classes who didn't wish to question established values, Freud, who trained as a physician in the science-dominated environment of the late nineteenth century in Vienna, sought to question everything. He always asked, 'Why?' Why do people fall sick? How do people get well? Why do we have wars? Why is civilization so often a failure? Why are the Jews persecuted?"[7]

Joyce went on to point out that psychoanalysis is an outgrowth of Western civilization and has had a vast impact on the Western world, particularly on the mental health professions. After Freud, all the healing professions began to think of bodily illness as connected with the mind. Freud was always concerned with understanding the links between psyche and soma. He was very aware that every bodily state has an effect on the images in the mind and that nothing happens in the mind or the psyche that does not also affect the body. He saw body and psyche as intimately connected, but governed by different laws of functioning. The laws of psychic functioning were not the laws of biological systems, but they constantly interacted and influenced each other.

"Freud's impact on the Western world has gone well beyond the mental health field. It has left a massive imprint on the teaching professions and has also had a considerable effect on many fields of creativity. Artists and philosophers in particular were very much inspired by Freudian philosophy and discoveries." Joyce paused and then added an afterthought: "Perhaps Freud has not had much effect on music. He claimed that the charms of music totally escaped him and regretted that this should be a closed world for him. However, he was passionately interested in words and language. It was most important to him to find words for human phenomena that had not yet been named. It could be said that Freud worshiped words. While it is certain that humankind is made by words and enslaved by words, much that is vital to human existence is also infra-verbal. Words, in a sense, represent the father, the external world. In both the Hebrew and the Christian Bibles we read: 'In the beginning was the word.' Might this be the heritage of a paternalistic religion? In any case I would like to suggest that in the beginning was the voice, and even in the intrauterine world the baby already hears sound and rhythm (perhaps the dawn of music?)."

A Topography of the Mind

Now Joyce introduced us to one of Freud's basic theoretical terms. "He was looking for a word to express the life force that is inborn in every human being—a force that invests life with meaning and seeks to reach out and touch other people; a force that finds expression in love, sexuality, religious feeling, and all forms of creativity. He thought of this force as a stream of energy and named it *libido*. But he also came to believe, as a result of his years of clinical observation and reflection on the world around him, that in the human being there is another, equally strong force that seeks death—self-destruction, and/or the destruction of others—and that there was an eternal conflict between the life force and the death force (*mortido*) in the human psyche. However, he came to conceive of the death impulses as originating from the libido. In other words this powerful source of life could be used for good or for evil—on the side of life, or on the side of destruction and death."

Freud employed some twenty-five different models of the mind. Joyce was not going to try to explain them all during her presentation, but instead would focus on some of the important ones, such as the theory of the life and death instincts and his model of the way memories are stored and the way knowledge is structured in the psyche. "Freud saw psychic structure as three layers. The top or cortex level he called *consciousness*. Next was knowledge of which we are not conscious at all times but which can always be recalled—this he named the *preconscious*. The third layer, the largest and most mysterious, is the *unconscious*, that which we do not know and cannot find in our waking, conscious lives and yet which exerts a vast influence on our behavior throughout life.

"The unconscious mind is constantly active in our inner psychic world, and drives us to find solutions to instinctual drives (which are often in conflict with the demands of the external world). The unconscious mind, Freud would say, is all of humanity—all that we have inherited from centuries of humanity. He called this our *phylogenetic* heritage, as opposed to our *ontogenetic* heritage, which is made up of all that one person has experienced from the time of birth. (Modern psychoanalytic research goes further and demonstrates the importance of events in fetal memory.) Thus memories from early childhood and even from the womb, along with the vital forces of libido and mortido, are all contained in the unconscious mind."

Dreaming and the Unconscious

After introducing these basic ideas from psychoanalysis, Joyce quickly turned to the issue at hand: dreaming. "This topographic model of the mind is important for understanding Freud's theories of how and why we make dreams, how and why we fall asleep—or are unable to sleep.

"The first puzzle to which Freud turned his attention is our sense of time. Time when we are sleeping is very different from time when we are awake. The unconscious, he says, is timeless, and it is while we are sleeping and dreaming that the unconscious finds its most direct expression—a vast everything-and-nothingness, which we can reach only with difficulty in our waking lives. When somebody tells us about a dream, he always says, 'I had a dream.' He never says, 'I am dreaming.' In this sense alone, the dream is always 'out of time.' We may even have what appears to be a repetitive dream, but it is never the identical dream as before, any more than an event that occurs more than once is the same event. We might therefore say that every dream, whether remembered or not, is an important event.

"Only two of Freud's twenty-three volumes of published work deal with the phenomena of sleeping and dreaming, and his major work on this topic, *The Interpretation of Dreams*, was essentially completed in 1896. Yet Freud continued to work on the innumerable groundbreaking ideas it contained for some thirty years to come. Freud himself considered *The Interpretation of Dreams* to be his most important contribution to the understanding of the human psyche. In fact it was from his study of dreams that he developed his whole theory of mind. In a discussion of psychic functioning he states that a person who is dreaming is not truly 'asleep,' even if he's sleeping. Although this sounded strange to those unacquainted with neurobiological research into sleep patterns, Freud already sensed that the mental state of sleeping was not the same as the dream state. He was creating concepts for what neurobiologists would discover fifty years later about REM and non-REM sleep. He had another hypothesis to the effect that when we are asleep or dreaming, our body is as though 'paralyzed' and that therefore dreams replace action. This was to him an important aspect of his exploration into why we dream."

The Dalai Lama had followed the account attentively, and for the first time needed some clarification. As usual, his question put the finger on a particularly slippery point. "If dreams replace action, in what way does one substitute for the other? Is one just happening while the other is not happening? Why do you use the word *replace*?"

Joyce replied, "When we are dreaming instead of doing something, we are existing in another state: a state of being without using *outward* action, that is, motivated physical action. When dreaming, although we are not moving bodily or reacting to events in the outer world, something very active is happening. A special process is taking place in the mind. Freud felt this was closely connected to the body. In fact, Freud couldn't quite work out why we don't dream all the time."

"So in some sense the body does have certain actions to perform in order to dream?" he questioned further.

"Yes, indeed. Freud proposed that all the dream thoughts and images that invade the mind are dealing with messages from the body. From there on he developed the idea that dreams are always connected with wishes (frequently stemming from instinctual bodily impulses), and that dreams were a way of fulfilling wishes. But what were these wishes? He said this begins with the simple wish to sleep and therefore with the need to let go of the external world. Then he posited a wish to stay sleeping, so that when thoughts and wishes coming from the unconscious mind caused conflict, in order to go on sleeping, we make dreams."

Like most of us unfamiliar with the Freudian clinical details, the Dalai Lama's interest was piqued by the model proposed. He asked: "The unconscious, the preconscious, and the conscious: all of these are providing the impetus to dream? On the other hand you've also said that the body sends messages to the mind during sleep. How do these interrelate? Are you saying that all these come from the body?"

"This touches the dynamic core of what the unconscious, as Freud conceived of it, has as its goal," Joyce clarified. She added that in fact the body is deeply involved in the unconscious. "Freud called instinctual impulses 'messengers from the body to the mind' such as 'I need love,' 'I am angry,' 'I am hungry,' 'I am frightened,' and so on. In this sense it is difficult to distinguish between that which comes from the body from that which originates in the unconscious mind. If the messages from the unconscious and the preconscious threaten to wake the person up, then one of the primary functions of dreaming was to prevent this. This concept led him to call dreams the 'guardian of sleep.'"

A moment of silence fell on the room while we contemplated the beauty of this phrase. The Dalai Lama proceeded with his questions, in his rigorous style of pursuing exact distinctions: "When you say that the preconscious is also an impetus for dreams, are you also lodging this in the body?"

"In both the body and the mind," continued Joyce. "Although the preconscious contains memories that can be recalled, these receive added importance when they are joined to libidinal bodily demands. Our conscious minds are bombarded daily with thousands of perceptions from the outside world as well as fleeting thoughts and feelings (more than we can give attention to—otherwise we wouldn't be able to go about the business of living). So we put these aside into those parts of the mind that hold recent memory, and these will often form the core of a dream that same night. Perceptions registered during the day, but to which we do not pay attention, are all the more likely to be used as the furniture of dreams when they are connected to bodily sensations, or to strong emotions. (Emotions are both a physical and a mental phenomenon.) Freud called such events the 'day residues.' Thus messages stemming from both somatic and psychic sources are used to make images, which will be woven into a story that becomes the dream. To answer your question, Your Holiness, we might say that the unconscious has a way of getting the sleeping mind to listen to the body through the preconscious."

"Freud asserted that the unconscious can never be known directly, and that we come closest to knowing this unknowable through dreams—as well as in certain states of psychic illness. People suffering from psychosis are also using parts of their unconscious minds to create hallucinations and delusions. And I would add that people who fall physically ill for *psychological* reasons are also using unconscious ways of letting the body talk. Furthermore, creative artists—painters, writers, musicians, scientific innovators, and so on—are also transforming and creating with messages from their unconscious minds. With regard to dreams, the unconscious uses the preconscious largely through the link with words."

"Then comes the complicated issue of what Freud called the apparently 'senseless and contradictory character' of dreams. He referred to the process of gathering up all the messages, day residues, and other factors that will be woven into a story of some kind as the dream work. He emphasized that it was hard work to produce the amazing phenomenon that we call a dream."

Narcissism

"Another important part of his theory of dreams concerns the question as to how an individual falls asleep. The libido can be oriented toward other people as well as being oriented to caring for one's own self and body. Freud called this latter investment of libidinal energy 'narcissistic libido.' It can be a healthy phenomenon, but it can also be pathological." In reply to a question from His Holiness, Joyce went on to explain, "The term *narcissism* is taken from the Greek myth of a boy called Narcissus, who fell in love with his own image and spent so long gazing at his reflection in a pool that he died beside it. The myth suggests that total narcissism would be the equivalent of death. But if we want to go to sleep, we have to be a little bit like Narcissus. We let go of our attachment to the outer world, to all the people who matter to us, to all the things that have happened during the day—we might say that we take them all back into our body-mind. This detachment from the external world indicates that libido must now become totally narcissistic, totally invested in the person to the exclusion of all other mental or physical occupations, if sleep is to ensue. Freud postulated that there was a regression to a state of 'primary narcissism,' and compared it to the mental state of a baby in the womb. He was not seeking to develop a *biological* theory about sleep, but a *psychological* theory to demonstrate the importance of instincts (which link body and mind together) in the sleeping state."

The Buddhist philosopher was intrigued by these last remarks.

"Isn't this narcissistic tendency at the point of sleep quite distinct from the narcissistic libido involved in self-absorption? The latter is quite intentional, whereas the narcissistic withdrawal in the process of sleep is purely natural, without intent."

"Yes, deeply natural and inherited over centuries. It is not intentional in the sense that being preoccupied with oneself narcissistically when awake could be seen as intentional," Joyce replied without hesitation. "We go back to an archaic narcissistic, or womblike state when we fall asleep and we seem to be very happy to stay there—but something forces us out of it when we begin dreaming. Today we would say a REM neural state provides the form of functioning in which dreams may most readily take place, but Freud held that unconscious and preconscious messages cause conflict and so we are obliged to dream in order not to be awakened."

Dreams, the Royal Road to the Unconscious

"Another important point related to the notion that the dream work is very active, and harder than daytime activity, is Freud's insistence that the psyche seeks to weave everything into one story, or into a unified set of images that are chosen to represent complicated events. Perhaps this story covers one's whole life; perhaps it is stimulated by thoughts and feelings from the waking state, and so on. Dreams often appear to be an attempt to find a solution to everyday conflictual situations in a person's life. A dream is therefore a story in disguise, and may include one or all of these elements.

"I will now discuss Freud's research path into the hidden meaning of dreams. Studying people under hypnosis first gave rise to some of his basic concepts about unconscious processes. Through hypnosis he discovered that the unconscious mind contains many memories that had been pushed beyond the preconscious—often events, thoughts, fantasies that we did not wish to remember. Under hypnosis, just as in dreams, these thoughts may come to light once again. (Even though modern research demonstrates that events recalled under hypnosis may be the result of suggestion rather than reality, this does not invalidate most of

Freud's theoretical constructions deduced from this field.)"

"Can you give an example of an experience which you want to forget and therefore you push it down?" requested His Holiness.

"Yes. I'm thinking of one of my analysands who, out of envy, had been very unkind to a friend and didn't want to remember it. That night, from these preconscious elements, he was led to remembering other unkind and envious actions from his distant past, forgotten things that came from the deeper unconscious. He had forgotten that out of sibling jealousy he had once pushed his little brother (whom, in fact, he loved dearly) off his tricycle, wounding him badly. The event had become an unconscious memory. But that night this man dreamed, not that he was pushing his colleague with hard words or being brutal to his little brother, but that an angry tiger was chasing a poor little dog. In the dream, he tries to fight the tiger and protect the little animal, but the tiger gets stronger and stronger—and eventually the man woke up in a panic. The meaning is disguised but becomes clearer when the patient begins to associate freely in the analytic session, around the different parts of his dream.

"The dream that a patient recounts in a session is what Freud referred to as the *manifest* dream, that which is evident on the surface, but his major interest was in the *latent* dream content, that is, the hidden meaning underlying the images in which the different themes struggling for representation are trying to find expression.

"Although Freud called dreams the royal road to the unconscious, he insisted that we can only scratch the surface of the unconscious mind. Most of its contents we will never know. However, he thought that dream analysis was a much surer method than hypnosis for achieving self-knowledge. In hypnosis people may remember things that are long forgotten, but when they wake up and you say: 'You pushed your little brother off the tricycle when he was only two,' they may say, 'Oh, did I? How interesting,' but they don't remember it or necessarily believe it. Freud concluded that just making something conscious because it came out while under hypnosis was not real knowledge in the sense that it carried conviction for the patient. True knowledge of the self, and all that one does not want to know about oneself, is in this way better attained through the experience of analyzing dreams. All the associations that come to mind, whether acceptable or unacceptable, whether recent or far back in the individual's life, coupled with the analysand's own reflections about him or herself, contribute to discovering the unconscious dimensions of the hidden self.

"Freud also spent many years studying his own dreams. Many of these are included in his books. He tried through his dreams to understand certain of his phobias, as well as jealous and angry feelings that he didn't wish to acknowledge until a dream forced him to become aware of them. He insisted that all analysts must continue to analyze their own dreams so that they could come closer to knowing many unpalatable truths about themselves which, if they went unanalyzed, might jeopardize their work with their patients. As for the importance of analyzing, sometimes for many sessions, every element in a patient's dreams, he concluded before long that this was not necessary. Patients should listen to their dreams, and associate to them if they so wished."

In a sudden burst of laughter, the Dalai Lama slapped his knee and joked: "It looks like there's an awful lot of work. If you had to analyze all your dreams there would be no time left to dream." We all joined in laughter, and Joyce added, smiling, "Indeed, it's a lot of work. Analysts never stop trying to get closer to their own psychic truth. But they don't stop dreaming." As she was saying this we all noticed that His Holiness had turned his attention elsewhere, and was using the conference brochure to help a little insect get to a safer place than the center of the coffee table!

Joyce continued, "The method Freud used for getting to the underlying truth concealed in dreams consisted in taking different parts of the dream and encouraging his analysands to use *free association*—to say whatever came spontaneously into their minds—around any of the elements of the dream. The overall idea was that you enter a state of un-integration, opening out, removing control, in which you are no longer hanging on to your cortical thinking, but instead letting ideas, perceptions, memories, and visions come up freely even when they appear incoherent, unconnected, or unacceptable. He found this method helpful for enabling his patients to discover the way the mind works—how their deep unconscious truths emerged through using preconscious verbal links to create the dream theme. Then, hopefully, they would be stimulated to apply conscious reasoning to reveal the hidden significance of their dream.

"To illustrate this idea of linking unconscious memories to a memory of something that happened the day before, let's go back to the patient who dreamed about the tiger. After recounting the dream he suddenly said, 'I don't know why, but that reminds me that yesterday I had a very angry discussion with one of my colleagues. I felt bad about it afterwards. After all he's a young colleague and looks up to me—but he sometimes says such stupid things! I shouldn't have been so unkind. . . . I'm thinking right now about my little brother who was eighteen months younger than me. I really loved him and we used to have such fun. Oh dear! I suddenly remember one holiday. . . . He was riding down the path on his tricycle and I pushed him over—he scratched both his knees and his chin had a huge bleeding gash and he started to cry. How horrible! Why did I do that to Bobby?' When I asked whether after Bobby's birth he was unhappy not to be the only one any more, he said, 'Yeah! I had to share *everything* with him. But then I loved him so much it didn't matter.' 'You loved him so much you pushed him off the tricycle?' 'Well . . . I guess I was angry with my mother. She was always busy, and then she got pregnant with my sister. . . . I don't know why she had to have three children!' That's called free association. As you can see, the only place we could ever allow ourselves to talk in that uncontrolled way is in psychoanalysis. If we did it elsewhere, we'd soon have no friends!" We all laughed at the image of carrying on life in free association, upsetting everybody with our conscious, or unconscious, ambivalence.

Joyce continued, "Free association leads you to express feelings and ideas that you've never wanted to tell anybody—not even yourself. Thus you get to a certain level of truth about your own self and your way of relating to others and to life in general."

I could see that the Dalai Lama was getting curious about these ideas. He interrupted the flow: "Are there any correlates in neuroscience to the three states—conscious, preconscious, and the unconscious?"

Joyce fired back without hesitation, and with a smile, "None whatever, I would think."

In my role as the coordinator it was up to me to put things into context. "In neuroscience the notion of an unconscious doesn't come up in any clear sense. Some people might say that it is related to the brain stem, the part of the brain that we share with older vertebrates, the reptiles, and which might have something to do with instinct. But it's a very vague comparison because the unconscious is also quite intelligent. The relationship is not really currently accepted. It's as if psychoanalysis and neuroscience are two independent streams in Western culture."

Joyce added that there is nevertheless some common ground. The neurobiologists have one set of theories about causes, and the psychoanalysts have other theories of causality, but they do complement each other. "Nobody can claim to hold the only or absolute key to the truth." She then moved her presentation from theory to clinical experience. "I hope I made it clear that the unconscious, the preconscious, and the conscious minds are always linked and interacting with each other, not only in dreams but also in waking life. People were shocked when Freud declared, 'We are not masters in our own house. We think we know why we do what we do, we think we know who we are and what we feel—but we actually don't know very much—we only see the tip of the iceberg.' People took Freud to task for proclaiming that human beings were not basically 'good,' that the forces of anger, murderous rage, and hate (not to mention sexual desires) are very strong right from the beginning in small children. The culture of the nineteenth century saw children as innocent—pure light—as though they were devoid of the impulses that, while they have to be controlled, are vital to human beings: love, hate, and incestuous and death-dealing wishes. This was shocking news and, although as a result he was attacked and persecuted in many ways, he went courageously on.

"I think that, apart from his brilliant inquiring mind, his determination to continue in spite of public attack may have been a way for Freud himself to deal also with his own suffering. In his earliest years his father and some of his best friends died. Then his half brother, whom he loved very much, and his daughter Sophie were killed in the First World War. Later his professional hopes of becoming a great neurologist were dashed. Then his closest follower, Carl Jung, whom he loved like a son, left him after some years of devoted collaboration. Some years later came the Holocaust and the Hitler regime. Freud's own family was threatened with death. But Freud went doggedly on with his momentous work. He was saved from the Holocaust and brought to London where he lived until his death from a painful cancer, in 1937. I think that, in addition to his scientific interest in dreams, the truth that he discovered through dream research comforted him and helped to maintain his vitality and his humanism, in spite of all of these tragedies."

Marie-Josée's Story

The mood of the room had become one of quiet reflection on the many ideas that had been evoked. But we all felt a bit lost in these flashes of Freud's personal struggles. So it was only appropriate that Joyce embarked now on an example that illustrated analysis in actual practice.

"I thought you might be interested in the dream of one of my analysands and the way in which her dream started off a dream of my own. I chose this to illustrate Freud's ideas on sleeping and dreaming, as well as to give a glimpse into the psychoanalytic process. I shall restrict myself to talking only about the small part of this patient's analysis that applies to our topic. 'Marie-Josée,' the name that we will use for the sake of privacy, was thirty-five years old when she first consulted me because, she said, she was having trouble sleeping. She was terrified at nights, unable to sleep without heavy doses of sleeping pills—but only when she was alone. Her husband, whom she loved dearly, traveled a lot, so she was often alone. Many times she also went back to her parents' home to sleep. But she felt this wasn't right at her age. She also suffered from agoraphobia and claustrophobia so that she had to avoid wide open spaces as well as small enclosed ones. She loved going to concerts, but she had to get a seat near the exit in case she felt 'closed in.' When she went to the hairdresser, she had to park her car where she could see it and keep the key close at hand so that, if she suddenly got frightened, she could jump into her car and go home. These symptoms caused her great mental pain, and she wanted to know why she suffered in this way, and what it meant. At our second preliminary interview, after giving me more details of her family history, she mentioned in an offhand way, 'There's another little problem, but it's not really a problem.' (I said to myself: Is this the real problem?) She had to urinate many times a day. She had been to two urologists who said there was nothing wrong with her physically. She added, 'It's not a psychological problem, it's just that my bladder is much smaller than other women's bladders.' After she left I wrote in my notes: 'Does she think she has a little girl's bladder and not a grown woman's?'

"During the first two years of analysis she hardly ever referred to her urinary problem. She would mention, for example, that her anticipation of going to a wonderful opera was ruined, for terror that she might not get the seat on the aisle so she could go to the toilet several times during the performance. She felt it was a problem that could not be solved, whereas her insomnia could be changed. Therefore I heard a lot about her insomnia, and little by little I got her to say what she imagined when she couldn't fall asleep. She said she was frightened that a man would come through the window and try to rape her. She would resist, of course, and he would kill her. I asked who she thought this man might be, but she could not work that out, or explain why he should be there whenever she was alone. I said, 'He's one of your characters; you have thought him up and put him outside your window.' She disagreed and insisted that this type of thing happened all the time, and she would bring me newspaper clippings about women being attacked by men, although she never found one in which a man climbed through a woman's window to rape and kill her.

"Finally, in order to get her to analyze her phobic invention, I told her a joke about a woman dreaming that a tall handsome man is approaching her. The woman screams out, 'What are you going to do to me?' The man replies, 'I don't know yet, ma'am—it's *your* dream!' She was able to laugh for the very first time about her rapist-killer, and said, 'Oh dear! It *is* my story!' As time went on, the fantasy slowly became an *erotic* one: 'I went to sleep thinking about the killer coming in the window and kissing me and making love to me.' (Eroticization is a most efficient way of overcoming many terrifying experiences and fantasies!) Eventually, Marie-Josée gave up the sleeping pills, but instead she now had to masturbate in order to fall asleep. It bothered her that she felt she had to do so whether she wanted to or not.

"Her other main topic was her mother. 'She rings me up all the time, always inviting me to concerts, always trying to get me back home. She's terrible, she just will not leave me alone!' bewailed Marie-Josée. We talked a great deal about this, and I had the impression that Marie-Josée's husband looked after her a little like a mother. She had another recurring preoccupation which she didn't think of as a problem—she did not want children. She was still a little child in some ways herself, and it seemed to me she thought there could only be one mother, *her* mother; she had to remain the little child with a little girl's bladder.

"One day she was quite angry with me. She said, 'I suppose you're very pleased that I can now sleep easily, but my day problems are just as bad as ever and my mother bothers me just as much.' I said, 'Perhaps I'm a bad mother to you because I haven't helped you solve your problems.' 'Yes,' she said. 'You're not helping me enough.' I asked her to tell me more about what she was feeling and she said, 'Yesterday I went to visit Suzanne, an elderly friend of my mother's whom I love very much. There was no parking place so I *had* to try to drive my car close to her house, because I was too terrified to cross the empty boulevard which was the only way to get near her street. I circled round for half an hour looking for a parking place. You see, I'm just as sick as ever! Then, I had to figure out how to get to my friend's house, because it's a one-way street. So I had a wonderful idea: I drove across the boulevard and I went *backwards* up the one-way street, and parked right in front of Suzanne's house. She said, "You're rather late. I thought you weren't coming." I felt so ashamed because I couldn't tell her why.' That night she had a dream: 'I was in a great stormy ocean, and I was very afraid. The waves got bigger and bigger although I looked around and thought that the scenery was beautiful—but it was terrifying as well. I thought I was going to die, and I said to myself, I must find something to cling to or I will drown. I saw one of those posts that you tie boats to; I can't remember what they're called. I reached out to grasp it, and it was made of stone.' *Pierre* means 'stone' in French and her father's name is José-Pierre. So I thought, 'Is she clinging onto her father?' She went on: 'I woke up terrified. I think this has something to do with my mother.' Now in French *la mere* is 'mother' and *la mer* is the 'ocean,' so it's understandable that through the verbal links your mother may be represented in a dream image as the sea. Marie-Josée had the same associations, for she went on to say, 'There's nothing new in this dream. It's about my overwhelming mother—I just get panicked all the time because she's so smothering and possessive.' I asked her: 'What is this post whose name you can't remember?' The preconscious memory suddenly came through. 'I know, it's called *une bitte à amarrer.*' In French, *bitte* is a post that you tie a boat to, but *bite*, pronounced the same way, is a slang word for the penis. The *bitte à amarrer* which ties boats safely to shore in an angry sea may also represent her father, symbolized by his sexual organ, and represented by his name—as though she were clinging to a father symbol to protect her against the overwhelming mother. Her next associations nevertheless surprised me.

"She continued: 'I'm thinking about my father and the day I saw him in the bathroom. I saw his penis and I knew I shouldn't see it. I was very excited and very frightened that my mother would be angry with me.' So now she's beginning to think that the dream represents her mother being angry with her because of her childhood sexual excitement when she caught a glimpse of her father's penis.

"One of the paradoxes of the unconscious is that everything that you feel is done to you, you also feel you are doing to the other person. It's the *connection* between two people that is important. So we might wonder if the dream is saying that Marie-Josée wants to drown her mother. (After all, she is the dreamer—she has invented the dream with its violent theme.) Then I began to think about her urinary problem. Little children have many fantasies about their body's secretions and often imagine that this is how their parents have a sexual relationship: they share spit, or they exchange feces or urine with each other. So I began to ask myself whether there might be a link between her urinary frequency and her childhood fantasies. Also, children have two contradictory attitudes to their body products: one is that they are a gift—a way of loving. On the other hand they are also imagined as bad and damaging. Good urine is giving something to mother; bad urine is punishing her—(smothering her maybe, in an angry ocean?). These were my thoughts: maybe every time she urinates she's flushing her mother out of herself or drowning her in urine. But I didn't say any of this because she had not given any associations that would allow me to make such interpretations.

"That was the end of that session and I felt disappointed. In my notes I had written that we hadn't come to anything new. She was still angry with her overwhelming mother, and still longing for her father's comforting and protecting presence. We'd worked over these themes many times, so what was I not hearing? Although I did not realize it at once, I had overlooked the fact that the elderly friend, a mother figure—was not an angry, frightening mother but a beloved one; and also, that in order to reach her she had had to break the law, and back up a one-way street. That night, *I* had a dream using these daytime residues. This dream made such a strange impression on me that it woke me up in the middle of the night and I couldn't get back to sleep. Finally I wrote it down because it puzzled me so.

"I have to meet somebody in a part of Paris that I don't know very well, but which has the reputation of being somewhat dangerous. Everyone is getting in my way, while I'm shouting, 'Please, I have an appointment.' A door opens and an exotic-looking Oriental woman says, 'Come on in.' She's clad in shimmering silk. I look at her and she says, 'You're rather late, you know.' I'm embarrassed because I feel very bad about being late, so I reach out and stroke her exquisite silk dress, hoping she will forgive me. Then I suddenly realize that I cannot be forgiven unless I do everything this woman wants. I think she's going to touch me and hold me. I'm going to have some sort of erotic relationship with her and I'm convinced there's nothing I can do about it; I just have to let this exotic creature do whatever she wants with me. I'm so frightened that I wake up. The dream theme is clearly homosexual, and as far as I could remember I'd never had any such dream before. It occurred to me that my two analysts, both men, had never interpreted any homoerotic longings (perhaps because they were so totally unconscious that I didn't give them the clues?).

"Unable to understand why I had made this dream, I began to free-associate. I thought immediately of Marie-Josée's session, through the verbal link to the words of her mother's friend, 'You're rather late, you know.' This fragment from my preconscious mind had probably touched off unconscious ideas that were quite obscure. What was the link between my dream and Marie-Josée's visit to her beloved elderly friend? Then, I recalled that she could only get there by taking a forbidden route. Of course the friend was a mother figure—and everyone knows that it's forbidden to have a sexual relationship with one's mother. But who was the beautiful Oriental woman in my dream? Another preconscious memory popped into my mind—some six or seven years earlier I had seen a Chinese woman for a few weeks only, who had sought help because of several difficult relationships with women colleagues and friends. I recalled that her father had three wives: the chief wife, a second wife, and the third who was her mother. The first wife was 'the real one who ran everything,' whereas she had complained that her own mother was 'more like a sister than a mother' to her. They would whisper secrets, talking about the father and the first wife, like children playing games.

"This is all I remembered of that patient's problems. I had been sympathetic over her sadness at not having a 'real mother' and having to accept a sort of sister-mother instead. Why had it not occurred to me that it could be rather nice to have a mother who was also a sister, with whom you could play games and whisper secrets? A rather special mother-daughter tie. For some reason I kept trying to remember this patient's name, and suddenly it came back to me: Lili. Then came a clear light. My own mother's name is Lillian! This beautiful Oriental was surely a dream disguise for my mother. Had I secretly wished for a 'mother-sister'? It then occurred to me that in many ways my mother was quite the opposite of Marie-Josée's mother, and I suddenly realized I could be *jealous* of their relationship. Why didn't I have a mother like that—always ringing me up, asking me to come home for the weekend, inviting me to concerts? I was now giving free rein to my associations. My mother was very active. She played croquet and golf, took singing lessons, enjoyed cooking meals for the family, made pretty little dresses for my sister and myself, and worked devotedly for the church to which we belonged. Mother was always busy, and was never grabbing at us. We were free to visit friends, go to the cinema, play sports, and so on. My sister and I thought we were lucky to have such freedom, compared with some of our friends. Then, suddenly another memory: I was about six and my father and mother came in to kiss us good night because they were going to a concert. My mother wore a beautiful, shimmering dress of apricot shot silk—the kind of silk that the lady in the dream was wearing. My mother in no way resembles an exotic Oriental, but I'm sure when I was six I thought she was absolutely beautiful. I would have loved to stroke her apricot silk dress, and although I always believed that I wanted to go everywhere with my father, I think now that I must also have wished that my mother would choose *me* rather than my Dad. I, too, would have a little silk dress just like hers and we would go to the concert together and push my father out.

"With these new insights I was looking forward to Marie-Josée's next session. At the session, she said, 'My mother called up again, she wants me to go to a concert.' I said, 'You complain a lot about your mother, but you also insist on how much she wants you to be with her. Are you trying to show me that even if she annoys you, her devotion is very pleasing to you, too?' After a shocked silence she said, 'Yes . . . and I guess I never told you that I ring her as often as she rings me.' Then she started to cry and said, 'My mother rang me a few days ago and said she and my father were going away on holiday for a week. She said she hoped I wouldn't feel lonely and need to be with them. They wanted to go away on their own and for once didn't want to be worrying all the time about how I was doing, alone in Paris!' I asked if she thought her mother shouldn't go away. 'Yes, she should, but it's true, I also want to be with her more than I've admitted.' So my dream was beginning to help me hear what she and I had not been listening to: how much she also *wanted* a close relationship with her mother.

"In the weeks that followed we continued to explore the unconscious homosexual longings of Marie-Josée's little girl-self, and the many buried fantasies that had only found expression in her phobic symptoms. Her fierce phobias of empty spaces and closed places slowly diminished. She was happy when her husband was there, but she was also happy with her own thoughts when he was away. The symptom of urinating all the time continued, however, and I was now trying to grasp its meaning in her deep unconscious fantasy life."

I could see the Dalai Lama listening carefully during this case story, following it with a mixture of attention and the amazement of someone who is not used to thinking about the mind as a psychological object, let alone confronting the neurotic illnesses common in modern urban life. He said, pondering, "How do you account for the fact that her agoraphobia and claustrophobia vanished?"

Smiling, Joyce answered, "Many reasons, but perhaps the most important was that she no longer saw her mother as totally bad, nor did she fear that her mother would kill her because as a child she had found her father sexually exciting. She now realized that she loved her mother as much as she loved her father. So her phobic terror of a man coming in through the window (which symbolized coming into her own body) in order to kill her turned into a man who would love her—the father whom she loved. But this was, she thought, forbidden, so we were able to put together that the empty or closed spaces had unconsciously symbolized the smothering mother as well as a space in which her angry mother would surge out at any moment to punish her. These unconscious fantasies had now become conscious thoughts which subsequently appeared absurd to my patient, and thus they lost their power over her mind. She also realized that although her mother was very demanding she was also loving and caring toward her only daughter—and also that it was not forbidden to love one's father. Although Marie-Josée would still get angry with her mother, she knew that she loved her, too, and that it was quite all right to have contradictory feelings towards the same person. She had also learned by now that little children love and desire both of their parents. These give rise to 'oedipal longings' as well as 'primary homosexual' longings. Although we had still not worked through her childlike erotic feelings for her mother, her relationship with her and with others had become much simpler. She no longer needed to protect herself from the frightening fantasies that were the unconscious cause of her severe phobic symptoms.

"Many things occur during the analytic process besides the interpretation of dreams—some of which are never even put into words. To begin with, the psychoanalytic relationship has a unique aspect in that two people are working together to understand one of them, each using their minds and everything that they can learn about their own truth to help understand the other person's truth. This in itself is a curative relationship. But Marie-Josée still had her symptom of urinary frequency. I assumed there was something we had not yet been able to put into words, perhaps to do with her masturbation fantasies which had become compulsive in order for her to sleep. It occurred to me that I had never invited her to explore her autoerotic fantasy life. When she complained once again about the compulsive nature of her nightly masturbation, I pointed out that this had replaced the old terror of the rapist-killer who had turned into an erotic figure. Perhaps her fantasies might help us understand what lay behind the feeling of compulsion. She replied without hesitating, 'Oh, in my fantasy there are men and women all loving me and touching me.' Then she stopped and said, 'But there's something I don't want to tell you. It's a silly thing; I have a little device for cleaning teeth with a stream of water. I use this apparatus to excite myself sexually.' I asked her to tell me more about the apparatus, and she said that her mother had given it to her but she had never used it for cleaning her teeth. I asked, 'Perhaps in your fantasy you are a little girl making love with your mother?' She said she had never thought of that, but felt it was profoundly true. She then recalled that when she was little she used to wet the bed. This released many more preconscious memories from childhood: how her mother would wake her up and take her to the toilet so she wouldn't wet the bed. She would say, 'Good girl,' when she urinated in the toilet and then carry her back to bed. This was a very tender memory. Suddenly Marie-Josée exclaimed, 'I know! It's just this little girl in me who didn't want to know how much she loved her mother. I now understand what you tried to show me—that I needed to have erotic fantasies about my mother as a little girl, so that I could become a woman like her.' Her symptom of urinating all day long gradually decreased, although it would return at times when external circumstances made her anxious or angry.

"All this took about five years of hard work. Her sex life became more gratifying to her and she began for the first time to think that she would like to have a child. She also began to travel with her husband since she was no longer frightened of open spaces and had overcome her fear of flying over water. I felt that she was growing into womanhood in every way. She no longer believed that there could be only one mother, and she was no longer a frightened child with a little girl's sex and a little girl's bladder. All these ideas were there already, in the core of her dream of the great ocean storm. It had taken much psychoanalytic work to uncover the truth that the death-bearing ocean was also a storm of love, love for her father coupled with the wish for a childlike, merged relationship with her mother, in order to become a woman and a mother herself. The angry sea of her internal world was slowly becoming an 'ocean of wisdom.'"

Beyond Freud

Marie-Josée's recovery made a very engaging story, and after a pause Joyce was ready to wrap up her presentation. "It is fifty-seven years since Freud died, and many people have continued his research on dreams and sleep. Some researchers have extended his major concepts; others have criticized them as new clinical insights prompted new questions. With regard to Freud's dream theories, one of the earliest and most important critics was Geza Roheim, a psychoanalyst who was also an anthropologist. He used his psychoanalytic knowledge to better understand primitive society and anthropology. After research with the Australian aborigines, he insisted that psychoanalysts should learn more from anthropology and that anthropology and psychoanalysis could enrich each other. In his last book, *The Gates of the Dream*, he described how he had come to realize that the same visions recur in the dreams of all mankind—the 'eternal ones of the dream' that reappear not only in the Occident but in all civilizations.[8] He came to the conclusion that the key to understanding another culture was to understand its dreams.

"Following Freud's idea that in dreams you return to an early feeling of being one with the mother's body, Roheim added that this was also a death wish. In addition to his idea of the eternal struggle between the life and death instincts Freud had added another dimension, that beyond the desire for life there is also a wish for the *death of desire*—a wish to return to an inorganic state of being, which Freud called the *nirvāṇa principle.* Roheim interpreted this wish for nothingness as a propelling force to deep sleep that represented the longing to merge with the mother. But there is another force, Roheim said: the body comes awake with its unconscious and preconscious thrusting towards life. This, he believed, represented the father; and the body, far from being 'paralyzed,' became a phallus. He thought therefore that in the dreaming state there is a constant conflict between the longing to merge with the mother and the longing to identify with the father as a powerful phallic symbol. He deduced that these two opposing forces clashed and that this clash was the cause of dreams. He also proposed that dreaming provided both sexes with a vital source of masculine and feminine energy, in the service of living. Roheim's research then led him to challenge Freud's contention that dreams were composed of only visual images and he was critical of Freud's approach, which treated the dream as a text—a text that required special knowledge in order to decode it.

"It is true that when a patient talks about his dream he is reporting something like a reconstituted text, something that happened in another state of mind and in another time frame. It no longer contains all the vital elements that are experienced and that constitute dream life. As a result, in the analytic process one is seeking to interpret the hidden meaning, but the interpretation is not the dream. A number of analytic writers have criticized Freud's hermeneutic approach. However, Freud himself was the first to point out that *dreams were not made to be interpreted.*

"Other analytic writers have proposed that the life and death impulses are re-created *by* the dream, that the very process of dreaming *creates* libido, and not, as Freud held, that the dream is merely a vehicle for the expression of libido.

"Then there is the interesting question of people who appear to be unable to dream. They believe they just sleep and wake up, with no experience of having entered another world that joins time and timelessness together. We have learned much about the incapacity to dream from an analyst who was also a pediatrician, named D. W. Winnicott.[9] Winnicott spent his life studying children and the childlike part of adults. He observed that very young children can go to sleep peacefully as long as they have what he called their *transitional object* with them: a teddy bear, a special toy, or a piece of their mother's clothing, and this precious object enables them to move from one world to another. When children begin to talk, they usually do not need their transitional objects any more. When they can say 'Mommy,' and can think about her reassuring presence, then language replaces the object. Winnicott thought of the transitional object as one of the earliest ways of creating a *space*, between oneself and the Other. This, he claimed, is the space in which creativity, art, religion, and all other cultural acquisitions come into being. These are in turn linked to what Winnicott called the *true self*—a dimension of the self that leads individuals to feel renewed, alive, in close contact with their own and other people's inner reality. He further advanced the notion that if there is no transitional space (for this space implies the capacity to distinguish between self and nonself) then this lack may inhibit, among other consequences, the *ability to create dreams*. Although everybody has a true self, for certain individuals who have suffered a lot in childhood, the true self is hidden beneath a false self. In this respect he criticized those analysts who interpret dreams all the time, saying, 'This means this and that means that,' because the patient runs the risk of giving the analyst what he thinks the analyst wants and this creates an even denser kind of false self in which the patient cannot 'let go' of rational thought because he is afraid of emptinesss. Winnicott thought of nothingness as a creative space in which you are receptive to new things growing inside your own mind, new light into the meaning of yourself and the world.

"It is time for me to conclude. We might resume by saying that dreams are the most intimate form of relationship we have with ourselves. In our dreams we return to our earliest objects of love and our earliest and most inarticulate conflicts with the difficulties of being human. As psychoanalysts we spend much time observing and trying to understand our analysands' reports of their dreams. So we must continually remind ourselves that we are touching something that is infinitely precious to the dreamer. We would do well to recall the lines of the Irish poet W. B. Yeats: 'Tread softly because you tread on my dreams.'"

Is There an Unconscious in Buddhist Teaching?

It was the end of the presentation time, and His Holiness acknowledged Joyce's effort with a big smile. She did not lose a moment in launching a question that was clearly burning for her, and for many of us: "I would like to ask Your Holiness if the Freudian concept of the unconscious has any corresponding ideas in Tibetan philosophy?"

He answered immediately. "First of all, within Tibetan Buddhism, you can speak of manifest versus latent states of consciousness. Beyond that, you can speak of latent propensities, or imprints (Skt. *vāsanā*; Tib. *bag chags*, pronounced *bakchak*). These are stored in the mind as a result of one's previous behavior and experience. Within the category of latent states of consciousness, there are states that can be aroused by conditions and others that are not aroused by conditions. Finally, it is said in Buddhist scriptures that during the daytime one accumulates some of these latent propensities through one's behavior and experiences, and these imprints that are stored in the mental continuum can be aroused, or made manifest, in dreams. This provides a relationship between daytime experience and dreams. There are certain types of latent propensities that can manifest in different ways, for example by affecting one's behavior, but they cannot be consciously recalled.

"However, there are divergent views within Tibetan Buddhism, and some schools maintain that these types of latent propensities can be recalled. This issue comes up particularly in relation to the topic of mental obstructions (Skt. *āvaraṇa*; Tib. *sgrib pa*), specifically obstructions to knowledge (Skt. *jñeyāvaraṇa*; Tib. *shes bya'i sgrib pa*). There are two categories of obstructions: afflictive obstructions (Skt. *kleśāvaraṇa*; Tib. *nyon mongs pa'i sgrib pa*) and obstructions to knowledge. Afflictive obstructions include such mental afflictions (Skt. *kleśa*; Tib. *nyon mongs*) as confusion, anger, attachment, and the like. Afflicted intelligence also falls into this category, for intelligence itself is not necessarily wholesome. It may be unwholesome; it can be afflicted.

"Concerning obstructions to knowledge, one school of thought maintains that these obstructions never manifest in consciousness; they always remain as latent propensities. Even within the Prāsaṅgika Madhyamaka philosophical school there are two positions. One maintains that all the obstructions to knowledge are never manifest conscious states, but are always latent propensities. However, there is a divergent view which maintains that there can be certain forms of obstructions to knowledge that are manifest conscious states.

"In one Madhyamaka text a distinction is made between recollection and certain types of activation of these propensities. A recollection is in some sense like a reenactment of the perceptual act you have performed; there is also an activation of these propensities that is not a recollection. The example given in this text takes the case of seeing an attractive woman while one is in the waking state, being attracted to her, but without taking much notice of her. Then in one's dream the woman comes to mind. This recollection is contrasted to the standard kind of recollection, because it arises purely through the stimulation of latent propensities. The propensities are stimulated and they manifest in dreams, and this process is quite unlike the process of straightforward recollection. There is another case, too, involving propensities: you engage in a certain type of action, be it wholesome or unwholesome, and through that process propensities are accumulated in the mental continuum until such time as they come to fruition. Until that happens, they are not something that can be recollected."

Joyce spoke what was in everyone's mind when she said: "It's as complicated as Freud's theory of what starts off memories and dreams. I'm very interested in what you say about these imprints that the child brings with it. The notion of an imprint passed on through centuries of man fascinated Freud; he called this our phylogenetic inheritance. The research on fetal memory observes imprints from the time the baby is in the mother's womb. Are these in any way similar to the imprints you call *bakchak*, Your Holiness?"

"It's very interesting," His Holiness answered. "At first glance, it seems that the notion of phylogenetic inheritance is quite different from Buddhism, in which these propensities are seen as coming from previous lives, carried from one life to the next by the mental continuum. However, in one of the treatises of the famous Indian Buddhist philosopher Bhāvaviveka, he mentions that calves and many other mammals instinctively know where to go to suckle milk, and that this knowledge comes as the result of propensities carried over from previous lives. The Buddhist theory of latent propensities speaks of these predominantly in terms of mental activity as opposed to the physiological constitution of the being. However, there are a lot of impulses and instincts in us which are in some sense biological and very specific to the kind of body that we have.

"For instance, Buddhism classifies sentient beings into different realms of existence, and our human existence is included in the *desire realm.* In this realm the bodily constitutions of living beings are such that desire and attachment are dominant impulses. So these can be seen in some sense as biological in nature. There are other propensities that are also related to one's physical constitution. For instance, it is said that the Fifth Dalai Lama came from a family lineage of many great tantric masters who had visions and other mystical experiences. He had extraordinary experiences quite often. These may have been due, in part, to his genetic inheritance from his ancestors rather than to his own spiritual development. Very deep tantric practice not only transforms the mind but also, on a very subtle level, transforms the body. Imagine this trait being passed down from parent to child. It seems very possible that if your parents and earlier ancestors had transformed the subtle channels, centers, and vital energies of their bodies, your own body would also be somewhat modified because of the accomplishments of your ancestors.

"Moreover, in Buddhism the *external* environment is seen in some sense as a product of collective *karma.* Therefore, the existence of a flower, for instance, is related to the karmic forces of the beings who live in the environment of the flower. But as to why certain types of flowers need more water, while other types need less; why some kinds of flowers grow in a particular area; and why certain types of flowers have different colors and so on—these matters cannot be accounted for on the basis of karmic theory. They must be explained mainly on the basis of natural laws and biology. Similarly, an animal's propensity to eat meat or to eat plants is related only indirectly to karma, but is directly due to physical constitution. Recall the statement by Bhāvaviveka about the calves knowing where to go for milk. Such behavior, which we consider to be instinctual, does indeed come from karma. But that may not provide a complete explanation. There may be more influences involved than karma alone."

On the Complex Inheritance of Mental Tendencies

"This comes very close to certain psychoanalytic constructions as well as an ever increasing interest in the individual's transgenerational inheritance," interjected Joyce. "In the course of a long analysis people discover knowledge that they have taken in unconsciously concerning their grandparents and great-grandparents—often knowledge of events that no one has ever spoken about. There has been considerable research in this area on the psychic problems of the children of survivors of the Holocaust. The survivors' children, or grandchildren, reveal through their stories, drawings, and dreams knowledge of their grandparents' traumatic experiences to which they have had no verbal access. Perhaps these psychological imprints, transmitted through the subject's genealogy, may also resemble the *bakchak* imprints from karma from generations back? Then, of course, there are purely biological genetics that make you look like your parents, but often like a previous ancestor too."

I was concerned that nonscientific concepts of time and inheritance were getting tangled up with scientific usage. "You're suggesting," I ventured, "that psychoanalysts believe the mother passes on influences to the baby unintentionally. What I hear His Holiness saying is that something comes with the mindstream of the individual that does not come through contact with the parents. Would analysts accept the idea that something comes not through the rearing of the very young, nor from genetics, but from a long-term mindstream?"

Joyce answered, "Indeed, this comes close to Carl Jung's definition of the unconscious, but it is not a classical Freudian viewpoint. However, it may be linked with what Freud called the *unknowable* in the human mind, that which we will never know but which belongs to all humanity."

"Could you give a precise definition of the unknowable?" asked the Dalai Lama, always looking to clarify terms. This is a trademark of his training, which resembles that of an analytic philosopher's in the West in its quest for terminological sharpness.

"Let me give Freud's metaphor with regard to the unknowable in the process of dreaming. When the patient is trying to understand and reconstruct, through dreams, associations, and memories, it is like unwinding a skein of wool. You can undo so much of it, but in the middle is a knot you will never be able to see into, that keeps the whole skein of wool together. This he called the unknowable and felt it was undefinable." We could not fail to notice that this was hardly a definition, and it reflected the metaphorical, almost literary, style of psychoanalytic work.

His Holiness insisted, "I've heard you say, on the one hand there is the phylogenetic heritage, with a purely physiological basis, and I would like to clarify if there is also a mental basis. Are you saying that the child gets an inheritance also from the streams of consciousness of its two parents?" Joyce confirmed this: "From both parents, and also there are the other two aspects of lineages which are quite distinct: one purely physical, the other one mental." After pondering this rather surprising exchange, he continued, "We could make a distinction between the *grosser* level of mind and *subtle* mind. In terms of the gross mind, there could be a connection from the parents to the child if, for example, one or both of the parents has such very strong anger or attachment that physiological changes take place in their bodies as a result of these mental tendencies. In this case, the mind influences the body. Thereafter they produce a child, whose body is influenced by the parents' bodies. The child's body, produced by the parents' bodies, could then influence the child's state of mind so that it similarly experiences very strong anger or attachment. In this case you'd see a gross level of mind, be it anger or attachment, going from one generation to another. That's a possibility. It's not just a pure mind-to-mind relationship, but a mind-to-body and body-to-mind sequence."

"From the biologist's point of view," I insisted once again, "the only possible inheritance consists of a physiological and a morphological organism. The idea that we can inherit what our parents have learned is called a Lamarckian evolution, which standard biology sees as false. Instead, I can only inherit from my parents such things as constitution and features; anything more I learn as a young child in contact with my parents. Biologically, it's a misnomer to call that *inheritance*. The term *inheritance* is reserved for the structural parental lineage, which is only a predisposition for the imprints we acquire as early learning by being with our parents. In biology this is the difference between phylogeny (the genetic inheritance) and ontogeny, which is what I learn, once I start my life. It seems that in Buddhism the notion of mindstream is neither phylogenetic nor ontogenetic, but represents a different kind of lineage, because it comes from a transindividual mindstream. This doesn't make too much sense in current science. I just was wondering whether in psychoanalysis this third category, neither learning nor physiological inheritance, is acceptable?"

"To the extent that character traits, such as being hot tempered and so forth, are thought to be inherited, would the biologist account for them in purely biological terms?" asked the Dalai Lama.

"This is a thorny issue, known as the nature-nurture debate. Most biologists would say that you can inherit certain tendencies for temperament, for example, but much of your actual temperament depends on the environment in which you are raised. You cannot reduce it to purely genetic factors or purely learning, because both are involved."

His Holiness continued, "Just to round off one point here: does biology refute the possibility that a person might have a disposition for anger which could influence their body; and then if that person has a child, the child's body would be influenced; and finally, the child's physical constitution would cause the child to have an angry disposition?"

I suggested that that would not be difficult to reformulate in biological terms. One could say, for example, that great stress or depression in a pregnant mother will physiologically affect the environment of the fetus, so much so that he or she will not be the same individual as if the mother had been normal. But biologists would call this ontogeny.

The conversation was getting sufficiently specific for others to contribute their views. Pete Engel continued with the biological track. "The whole debate over what is genetic and what is environmental has greatly changed in recent years because of studies of separated twins. There is now a belief that much more of what we consider as 'self' is inherited and much less is environmental than was previously thought in Western science. The studies have been done with identical twins who were separated at birth for various reasons, who may have grown up in different countries, with different parents, without knowing that they had a twin. When they were brought together, there were many more similarities than non-similarities in most cases, and many more than anybody had expected. They might, for instance, wear the same clothes or hairstyle, have the same jobs, or have married people with the same first name." His Holiness remarked that he had been aware of these studies, but that the similarities were not invariably present. Clearly, the studies had a bearing on the point, but did not prove that one could inherit traits that one's parents had learned.

The psychoanalyst rejoined the debate. "I'm not so sure," said Joyce. "We inherit, in preverbal ways, many character traits and tendencies that have marked our particular family's history and ways of reacting. Present-day psychoanalytic researchers claim that children are all born with what they call a core self, which may include characteristics that do not belong to the parents. They are not just blank screens with a genetic inheritance on which the parents are going to write the first structures of their minds. They already have minds of their own. Apart from fetal imprints, it seems to come from many generations back, though not from a mindstream of knowledge that has nothing to do with either the generations or ontogeny. But certain schools of analytic thought would be ready to accept the concept of the 'third category' of innate knowledge—particularly those of Jungian inspiration."

Foundation Consciousness and the Unconscious

The discussion of inheritance seemed to have continued long enough, and I was interested in returning to parallels with the unconscious in Buddhist theory. Anyone exposed to the corpus of Buddhist theories known as Abhidharma is familiar with the notion of the *ālayavijñāna*, usually translated as the *storehouse consciousness* or the *foundation consciousness*, an existential background from which all manifestations in daily experience seem to arise, and which is accessible to direct introspection during meditation. I was curious whether the *ālayavijñāna* might be related to the unconscious.

His Holiness's response was fascinating. "The very existence of this foundation consciousness is refuted in the Prāsaṅgika system, which is generally considered by Tibetans to be the highest philosophical system in Buddhism. In brief, this foundation consciousness, or storehouse consciousness, is believed to be the repository of all of the imprints or *bakchak*, the habits and latent propensities that one has accumulated in this and former lives. This consciousness is said to be morally neutral, neither virtuous nor nonvirtuous, and it is always the basis of latent propensities. Finally, it is nonascertaining; that is, it can have objects as its contents, but it does not ascertain, or apprehend them. Phenomena can appear to it, but it does not ascertain them. But the main difference between the foundation consciousness and the psychoanalytic unconscious is that the *ālayavijñāna* is manifest to consciousness. It is ever present, and it is manifest in the sense that it is the basis or core of the identity of the person. In contrast, the psychoanalytic unconscious is something that you cannot ascertain with ordinary waking consciousness. You can have access to it only through dreams, hypnosis, and the like. The unconscious is concealed, and what becomes manifest isn't the unconscious itself, but rather the latent imprints, or propensities, that are stored in the unconscious. On the other hand, what is stored in the foundation consciousness can become conscious, and the foundation consciousness itself is always present."

"So that's the basic distinction between the Freudian unconscious and the *ālayavijñāna*: the foundation consciousness can manifest without disguise, without going through a dream?" I asked.

"Yes, it's more like consciousness itself, because it's functioning as a full-blown consciousness all the time."

"From a psychoanalytic viewpoint a baby's first external reality is the biparental unconscious. Could this include their *ālayavijñāna*? Or the baby's?" asked Joyce.

"The foundation consciousness is regarded as a continuum coming from beginningless time, a stream of consciousness that is carried through successive lifetimes. In Buddhism reincarnation is generally accepted and according to the Yogācāra school of Buddhism, the foundation consciousness accounts for the transition from one life to another. Moreover, it's the basis on which the mental imprints are carried, in the newborn baby and in both parents separately."

The Yogācāra was an important school of Mahāyāna Buddhism that flourished in India beginning in the fourth century C.E. It was also known as the mind-only doctrine. Its proponents were idealists who held the view that no reality exists outside of consciousness, and who developed the theory of *ālayavijñāna* to account for the apparent coherence of phenomena. It is interesting to observe that in Buddhism, just as in psychology or biology, there are many conflicting interpretations. The Dalai Lama clarified his own stand: "As far as my own position is concerned, I totally refute the existence of the foundation consciousness. The reason that the Yogācāra school felt the need to posit such a category of consciousness was not because they had strong inferential grounds or experiential evidence indicating its existence. Rather they posit this out of desperation because they were philosophers who believed that phenomena must exist substantially. They wanted to believe that the self was findable under critical analysis. The self cannot be posited in terms of this continuum of the body, because the body ceases at the time of death; and this school affirms the idea of rebirth. So, in positing the self, they needed something which would carry on after death, and this has to be mental. If any other types of mental consciousness were to be posited as the self, then they could be either wholesome or unwholesome; and they could change through various stages. Moreover, there are also meditative experiences in which the individual remains in a nonconceptual state, during which all states of consciousness that are either wholesome or unwholesome cease to exist. Yet something has to carry on. For all these reasons the Yogācāra school posited the existence of an additional category of consciousness which was called the foundation consciousness. This step was taken on purely rational grounds. They were compelled to formulate this consciousness because of their rational presuppositions, rather than through empirical investigation or realization."

Imprints and the 'Mere-I'

"How the imprints, or *bakchak*, of the mindstream fit into this picture requires an examination of other schools in Buddhism." His Holiness was referring here to the schools of thought that preceded the Prāsaṅgika Madhyamaka school, which is considered to be the most philosophically advanced in the view of the Gelugpa monastic order to which the Dalai Lama belongs. (The Gelugpa order was founded by the reformer Tsongkhapa in the fifteenth century, has grown dominant in numbers over the years, and has as its visible spiritual leaders the lineage of the Dalai Lamas.) The Prāsaṅgika school is a product of the second wave of the development of Buddhism known as Madhyamaka, led by the great scholar Nāgārjuna (c. second century C.E.). An early school of particular historical interest is the Svātantrika Madhyamaka (prominent in the fifth century C.E.), based on the writings of the Indian adept Bhāvaviveka. "The Svātantrika school says that you don't need to posit a foundation consciousness. The continuum of mental consciousness itself will act as a repository for these imprints. This is where Bhāvaviveka leaves it. However, this position is also problematic, because there's a specific state along the path to enlightenment called *the uninterrupted state on the path of seeing*. In this state one passes into an utterly nonconceptual, transcendent awareness of ultimate reality; and it's said that this state is utterly free of any tainted consciousness. That being the case, this transcendent awareness is not a suitable repository for various wholesome and unwholesome imprints. But it's not clear that Bhāvaviveka ever raised that question or responded to it.

"Now we return to the Prāsaṅgika Madhyamaka perspective, which is a critique of all the preceding views, including those of Bhāvaviveka. In response to the previously mentioned problem, the Prāsaṅgika school says you don't need to posit even the continuum of mental consciousness as the repository of latent imprints. In fact all of these problems arise because of an underlying essentialist assumption that something must be findable under analysis, something that *is* the self. People come up with different ideas: the foundation consciousness, and the continuum of mental consciousness; but they are all vainly trying to find something that is essential, something that is identifiable under analysis. And that is the fundamental error. If you get rid of that error, as the Prāsaṅgikas do, then there is nothing that can be found under analysis to be the self. You give up that task altogether, and you posit the self as something that exists purely by conventional designation.

"Then we come back to the issue of the repository for these mental imprints. The Prāsaṅgikas also assert that if you engage in a certain action, you accrue certain mental imprints which, we can say for the time being, are stored in the stream of mental consciousness. You don't need to assert an internal substantial continuum that will act as a repository for these imprints forever and ever. You don't need to assert a substantial continuum of anything that is truly, or intrinsically, the real repository for all of the imprints. You don't need it because both the mental continuum and the stored imprints exist only conventionally, not substantially. For this reason, you don't have to worry about the case of the nonconceptual state of meditation. According to the Prāsaṅgika view, the imprints are stored in the *mere-I*. Now what is the nature of this mere-I? Where is it to be found? There is nothing really to be found; it's merely something that is designated in different ways. Returning to the problematic situation of the nonconceptual state of awareness of ultimate reality: at that time, upon what are these imprints placed? Upon the mere-I, because there's still a person there, purely as a convention. So that person is the repository, but this is not a repository that can be found under analysis, as these other schools assume.

"One can speak of the 'I' being designated on the basis of the gross aggregates (psychophysical constituents) or the subtle aggregates. Similarly, the 'I' can be designated on the basis of gross consciousness or subtle consciousness. One way of looking at this statement, that the mere-I is the repository of mental imprints, is to look at it from a conventional point of view. When a person has done an action that leaves certain imprints, he now has a certain propensity due to that experience. That's all there is to it. You don't need to posit a substantial basis that exists as the repository for that propensity. That is the Prāsaṅgika Madhyamaka view."

More on Mere Identities

This subtle and elaborate explanation of how identity appears in the Buddhist tradition could not fail to arouse the questioning of our resident Western philosopher. Charles Taylor struggled to reflect his understanding of the argument thus far: "Perhaps analogies with Western philosophy might help. Hume made a famous statement about the self: 'I look within myself and I try to find a particular item which is the self'; and he failed to find one. I think you are saying in part that he was asking the wrong question by trying to find a particular element that you could single out under analysis. But you could have another view of the self as something that presents itself as a self without any continuing element. The Western analogy is that of a ship: if you change a plank on it every year, at the end of so many years you could perfectly well say this is the same ship, even though all the pieces of wood are different. You can give a causal account of why this ship is a single, continuous, causal stream. I had assumed that a similar account of continuity over lives would explain the Buddhist view of how an imprint can operate across lives in this continuing entity. It's a continuing entity because it has a continuing causal history. I thought that was going to be the answer, but instead—"

"What is your view on the ontological status of universals as opposed to particulars?" interrupted the Dalai Lama. "As you know, *universals* and *particulars* are extremely dense terms in Western philosophy, and they are also very precisely defined in Buddhist philosophy. When I use these terms as translations from the Tibetan, their meanings are by no means guaranteed to correspond to the Western philosophical terms. For instance, in the case of this ship, let's first look at it in terms of its temporality or specificities. In this regard the ship of the first year is not the ship of the second year and so on; nevertheless, at the end of thirty years you can still talk of the ship as a generality (Skt. *sāmānya*; Tib. *spyi*). Would you accept that this generality of the ship could be called a universal?" Charles agreed that this was a universal. His Holiness continued, "All the time you have specific temporal instances of the ship: you have today's ship, A; tomorrow's ship, B; and the next day's ship, C; and A is not B and B is not C, and so forth. At the same time, there's a ship all along and that is the universal of the ship. You can say the ship of the first year is the ship, the ship of the second year is the ship, and so on, identifying the specific temporal instances with the universal. But if you ask whether this ship is the ship of the first year, the answer is no. So you cannot identify the universal as a specific temporal instance.

"Let's move to another Buddhist philosophical system now, the Sautrāntika system. According to this system, the person, or self, is neither a physical phenomenon nor a mental phenomenon, and yet it is an impermanent phenomenon subject to change. Physical phenomena are considered self-supportive; that is, they substantially exist. According to this system, you can really point your finger at them as something truly, substantially present. This is said to be true of the body, and also of nonphysical phenomena such as mental processes. The mind is really present, substantially existent. According to this system, in contrast to the two discussed earlier, the self does exist, but it does not exist substantially. This being the case, there would be some difference in the continuity of the self as opposed to the continuity of your ship, which is a purely physical phenomenon. In the Buddhist view of universals and specific instances, you can speak of 'I' as a youth, 'I' as a middle-aged person, and 'I' as an older person. Or, taking myself as an example, I am a human being, I am a monk, and I am a Tibetan. All of those are specific instances of 'I,' but then there is also the universal of the 'I.' The specific 'I's and the universal 'I' are not identical. They are distinct; yet they are said to be of the same nature. It is possible to view the 'I' as a universal, and then to examine its specific instances, but that does not mean that the specific instances and the universal have separate natures."

"Yes, they're just different descriptions," agreed Charles. Not completely satisfied with the original issue, he pursued it: "Can I come back to the question that still puzzles me? Supposing I have a *bakchak* or imprint of some kind. For example, I prompt anger, and you would say that many lives ago I did something which brought this about now. So you're making a causal attribution, a causal relation between something that happened a number of lives ago and my *bakchak* today. What kind of continuity must exist between the earlier event and myself now, in order to make that causal attribution?" I was glad that he posed this key question so clearly.

Gross and Subtle Mind

The answer came that in Tibetan Buddhism there are two perspectives on this: sūtra and tantra. The sūtra perspective comes straight out of the Buddha's teaching, and is considered the common basis for all Buddhist schools throughout the world. In contrast, the tantra perspective is a more secret and esoteric set of teachings that has been embodied in lineages of yogis and mystics from Tibet, and earlier from India. The tantric perspective is considered by most Tibetans as the highest. "I have been speaking from the Prāsaṅgika Madhyamaka perspective, which is still on the sūtra level. This view does not posit subtle mind and the clear light; so we have to account for this continuity of self, or person, without invoking the concept of the subtle body and subtle mind. From the Prāsaṅgika Madhyamaka point of view, the continuity of the *person* is maintained in an analogous way to your example of the ship. One person can have specific identities, such as 'I am a monk,' 'I am a yogi,' and so forth, while the universal 'person' is still applicable to all those identities. The validity of the person's continuity is explained conventionally: in conventional terms, you can validly say that I had a previous experience at a particular time which has resulted in my present behavior. That is, you can maintain that this person, 'I,' who is experiencing the consequence now is 'the same' as the person who had the earlier experience.

"The tantra, or Vajrayāna, perspective is perfectly compatible with the Prāsaṅgika view, but also posits something further, namely a continuum of a very subtle mind, and a continuum of very subtle vital energy, which is of the same nature as that subtle mind." This fundamental notion of very subtle mind or clear light was to reappear on numerous occasions in the days that followed. Here we had a first foretaste of this multifarious concept. "This twofold continuum is forever unbroken, from beginningless time to the endless future; and this is the subtle basis of designation for the self. So the self can be designated on the basis of gross physical and mental aggregates, and also on the basis of these very subtle phenomena. There are certain occasions when the very subtle vital energy and mind are manifest but the gross aggregates are not; and on those occasions the self is designated on the basis of those subtle phenomena. So you always have a basis of designation, either gross or subtle, for the self. For this reason you have continuity, even after one has become enlightened and liberated from the cycle of existence, at least according to the sūtra level of interpretation. In one text, Maitreya gives an analogy of rivers coming from all different directions, merging into a single ocean in which there are no distinct identities. However, that's not the position that Vajrayāna takes, and I would dispute the claim that even in the state of enlightenment the continuum remains."

He then turned to Charles with a fixed gaze. "So we come back to your question, Professor Taylor. We can ask whether the continuum of the very subtle energy-mind also exists purely conventionally, or does it have some kind of substantial existence unlike everything else? In fact, its existence is purely conventional, and this point is extremely important. Now you might ask: what designates the self on the basis of this very subtle energy-mind? Does it designate itself? It does not. When the very subtle energy-mind is manifest, it is nonconceptual. It's not the type of awareness that cognizes an object or conceptually designates anything whatsoever. It's technically described as a nonconceptual state. When this very subtle energy-mind manifests, one has no sense of a self, and *that* is the main point. When we speak of designating a self on the basis of the very subtle energy-mind, this is done from a third-person perspective, not from the first-person. We mustn't confuse the two! For example, when the very subtle energy-mind is manifest, it does not have the clear light as its object. It does not apprehend anything as an object. It itself is the clear light. Similarly when you abide in meditative equipoise, experiencing ultimate reality, you are not aware that you are in meditative equipoise. But if you are very well trained in such deep meditative insight, after such an experience you can look back on it and think, 'At that time I was experiencing the clear light.' This is now, in a sense, a third-person perspective. It's an outsider's point of view looking back on your own experience of meditative equipoise at a former time. But you certainly do not think anything at all while you're in that state itself. You're not thinking in terms of existence, nonexistence, or any other conceptual category.

"This very subtle energy-mind is considered subtle relative to gross phenomena, but that does not mean that it is therefore findable under analysis, or that it has some kind of a substantial, intrinsic existence. It does not. One could ask whether the continuum of gross consciousness, with its various mental processes, is distinct from the continuum of very subtle energy-mind. Do they have separate natures? The answer is no; they are not distinct continua with separate natures. Rather, there is an unbroken continuum of the very subtle energy-mind, and from this arise the grosser mental states."

Conventional Designation

Charles again tried to clarify, asking the Dalai Lama to define the word *conventional* in the sense that he was using it. His Holiness explained this important notion at length. "First of all, it can refer simply to ordinary human experience in which people quite spontaneously, without any special education or philosophical background, make such comments as, 'I went here; I did this; I am that; I am fat; I am skinny,' and so on: all the ways that we normally speak about ourselves, frequently using the word 'I.' In this context, I could assert that I exist for the sole reason that here I am talking.

"However, one may not be satisfied with that, but ask further what the nature of this 'I' really is, and start to seek it out. If something were findable under that type of analysis—if you could find the 'I' when you set aside conventional usage and asked what it really is—that would be something that exists not merely conventionally. The Prasaṅgika Madhyamaka system refutes the existence of any 'I' that is findable under analysis. Let's step back and review how we normally, or conventionally, speak of the self. We say, 'I am tall; I am this; I am that; I did this; I did that,' and we are satisfied to leave it on that level. In that conventional sense, I exist, quite in accordance with the way we normally speak. The self exists as we conventionally designate it, just as we normally speak about it. It exists in this fashion if we don't investigate its actual nature. Leaving aside the question of how it really exists, if you simply attend to ordinary, conventional usage of the term 'I,' then such statements as 'I am tall,' and so on are certainly true. One way of looking at this is to consider the scope of analysis. When we speak of the conventional nature of reality in the Buddhist sense, we are in some sense accepting certain limits to the validity of everyday discourse. For instance, the moment we step beyond the conventional usage of the term 'I' and start to ask about the true referent of this designation 'I'—asking, for example, what exactly it is that continues—we are taken beyond the boundary of everyday discourse. If you can find such a true referent of 'I,' then from the Madhyamaka perspective the self would be ultimately existent. Since that is not the case, existence can be understood only within this framework of conventionality. That's one way of looking at it.

"Now, if that were the case it would follow, absurdly, that anything that is designated can be said to exist. But although anything can exist by designation, that does not mean that anything that can be designated exists. In other words, it's not true that anything you dream up or conceive therefore exists, simply because you've designated it or thought of it. But anything that does exist, exists by the power of conceptual and/or verbal designation. When we speak of the existence of everyday objects, we use two criteria jointly. One criterion is that the designation is a publicly accepted convention; it is part of everyday discourse. But this does not imply that truth is determined by majority rule. Something is not more true if more people believe it, and less true if only a minority believes it. So, for an everyday object to be deemed existent, not only must it be accepted in everyday discourse; there must also be no facts in common experience that are incompatible with the existence of that object. These are the two criteria for the existence of everyday objects.

"When we speak of philosophical issues, then we have to add one more criterion: that ultimate analysis, which seeks the true nature of the object in question, should not negate it. Consider, for example, whether the foundation consciousness (*ālayavijñāna*) discussed earlier exists or not. How do we determine whether a statement that such a thing exists is true or false? The philosophical school that posits the existence of this consciousness does so on the assumption that there must be something essential which is the true person. This is a kind of ontological commitment to the idea that, when you search for the ultimate nature of a reality, analysis should not negate the existence of an object if it exists essentially. To the contrary, however, when you search for the ultimate nature of the foundation consciousness, it proves to be unfindable."

Charles persevered, "Are there things that exist, not conventionally but ultimately? Is there any way of describing what such an element might be?"

"There are two ways of understanding the term *ultimately existent* in the Madhyamaka school. In one case, it refers to a certain type of analysis that probes into the deeper nature of reality. For instance, let us take an object, like this microphone. You probe into the nature of the reality of this microphone, and what you find is the absence of its identity. There is no essence, no true referent, and that realization is reached through what we call 'ultimate' analysis. From that point of view, the 'identitylessness' of this object is ultimate. But this absence of identity would not be called 'ultimately existent.' Although the absence of identity may be seen as ultimate, it's not ultimately existent; because if you take that itself as the object of further analysis and search for its true nature, what you find is the identitylessness of identitylessness. And this goes on ad infinitum. Therefore even the absence of identity exists only conventionally.

"Now, there are in fact two ways of positing nonexistence. One is by determining that a posited designation conflicts with a valid convention such as the normal use of language. For example, if you were to say that this is an elephant here, obviously there's a lot of conflicting evidence about how we use that term. That's one way of determining the nonexistence of an elephant here. Another way of positing the nonexistence of something is by means of ultimate analysis.

"Further, there are three ways in which something can be negated. The first is by mere convention. For instance, someone may claim that this person is John, and you can dispute that statement by saying that he is Alan. In this case, the decisive factor is popular convention. The second way to negate the existence of something is by means of logical inference, and the third way is by direct perception. There is also a fourth way that entails reliance upon a higher authority, and this requires trust. For example, someone might claim that I was born in the year 1945. But I can dispute that by saying that I was born in 1935, and here the authority that I am invoking is the testimony I have heard from others. I don't know this all by myself, but I was told this by others who can speak about this with authority. Much scientific knowledge is accepted by nonscientists in this way, based upon testimony or authority. We don't know the truth of many scientific claims in that we haven't proven them, but we accept the authority and testimony of people we trust. For example, I accept the previously mentioned claim concerning REM sleep, based on your testimony."

Psychoanalysis as Science?

By this point, the philosophical precision was beginning to lose some of the participants, although the fundamental issue was clearly to move beyond simplistic answers about what Buddhism, or the West for that matter, understands by self, identity, and continuity. Charles was aware of this and said, smiling, "Perhaps we should try to get back to psychoanalysis. I'm sorry that perhaps these issues took us a little beyond that. I want to comment on the nature of psychoanalysis in relation to yesterday's agenda because this is disputed in the West. Some people say that psychoanalysis is not a science in the same way that neurophysiology is, because a criterion for a natural, hard science involves identifying its subject in a way that is independent from our own moral, spiritual, or emotional lives. The language in which people describe their moral aspirations and so on is very varied and always disputed. Western natural science succeeded because scientists found other ways of identifying and describing things that abstracted from these differences. Psychoanalysis is a science that uses these terms to deal with our moral, or spiritual, or emotional life: people's feelings, their sense of self, and so on. The result is that many people in the West see psychoanalysis as an interpretive science or hermeneutic science. A certain degree of controversy is accepted in this type of science that will never be fully resolved. New theories come along and sweep aside the older ones. You might say that Freud invented his own mythology when he talked about the life instinct and death instinct; someone else will talk about the father principle and the mother principle. These are competing mythologies and it's not very clear how different mythologies can all be serviceable in helping to cure someone. But psychoanalysts from different schools do have some success in curing their patients, though we don't fully understand how getting people to see their lives in these terms helps them.

"A second point is that all the different interpretations still fit very much within a Western outlook. Freud and all his followers think they're exploring the depths within. In other traditions, dreams might be read for purposes of divination, to see what's going to happen in the future. This would not be reading a dream to look inward but rather to see something beyond the dreamer. Because of all this, do you think the findings of psychoanalysis would hold up across different cultures? For instance, does prophetic dreaming exist in Tibetan culture?"

His Holiness responded, "From a Buddhist point of view one cannot expect to have a similar methodology for investigating the mind as you do for analyzing measurable physical phenomena, in which you can seek universal laws and uniformity, and from them derive scientific principles. In the case of the mind, the variables are so rich that even excluding the phenomenon of rebirth, within this lifetime alone, there are so many factors that cause the diversity of people's mental attitudes, inclinations, desires, interests, and so forth. The variables are so much more complex that you cannot expect to find uniformities and laws similar to those of physical phenomena. Because of the immense complexities of the human mind, and because of the great variability from one individual to another, all we can do is simply describe mental events once they have occurred. It becomes extremely difficult to make any universal statements that will always be true for everyone at all times."

Joyce concurred, "In other words, we are all working with and maybe creating different theories. But by definition a *theory* is merely a set of postulates that has never been proven—and may be forever incapable of proof. (If these postulates could be proven they would be laws!) In this sense, psychoanalysis is an anthropological science whose theories can never be proven. It is a coherent set of theories, supported by clinical observation, and therefore continually evolving."

This was a fitting point to stop, for it was already noon. The Dalai Lama saluted everybody warmly, and departed. After that all was quiet except for birds in the yard.

4

Lucid Dreaming

JAYNE GACKENBACH IS A PSYCHOLOGIST who does social science research, part of a small group of scientists worldwide who have been interested in sleep and states of consciousness for the last fifteen years.[10] She took the hot seat as the session started in the morning of an amazingly crisp and sunny day.

"My task here is to address studies of lucid dreaming, although later I will also introduce what I call 'witnessing dreaming.' A *lucid dream* is a dream in which one is actively aware of the fact that one is dreaming. In such a dream, where this awareness is separate from the content of the dream, one can even begin to manipulate the story and the characters to create a desired situation. For example, in an unpleasant dream situation, the dreamer might reflect, 'I don't have to put up with this,' and then change the dream or at least back out of the involvement. *Witnessing dreaming* is a dream in which you experience a quiet, peaceful inner awareness or wakefulness, completely separate from the dream.

"My colleagues who do this research have mostly concentrated on lucid dreaming as a function of self-reflection. In normal dreams, particularly those of young children, the dreamer does not usually appear in the dream. It is very important to note that lucid dreams can emerge out of any dream, and that emergence is self-reflection."

Evidence for Lucidity

"Reports of lucid dreaming exist in many cultures, and in Europe date back to the earliest periods of recorded history.[11] But instead of a long historical analysis, I want to address the status of lucidity in science today.

"A consensual scientific validation of lucidity did not in fact happen until the mid-1970s, when Keith Hearn and Allen Worseley in England, and Stephen LaBerge at Stanford simultaneously discovered a way to prove lucidity using electroencephalography.[12] Independently, they each came up with the same experiment: to ask the subjects to signal by moving their eyes when they enter lucid dreaming. Presumably, they postulated, moving the dream eyes would be reflected in movements of their bodily eyes, which can be, as you know, externally monitored. The beauty of the idea of course is that during dreaming all other muscle movements are blocked.

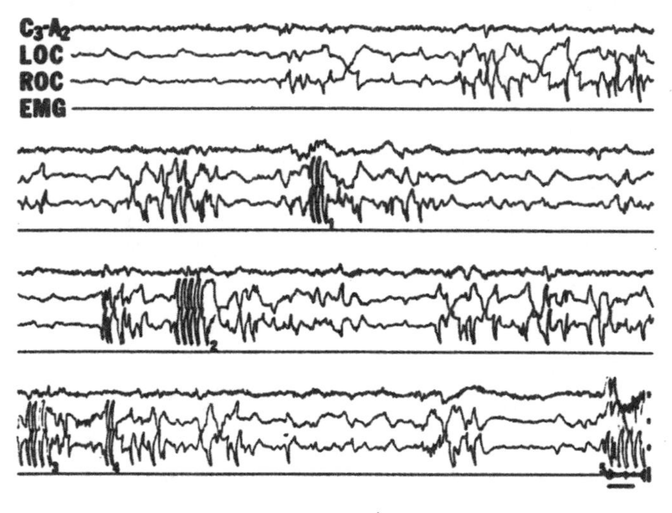

Figure 4.1

A typical dream-initiated lucid dream. Four channels of physiological data (central EEG [C3-A2], left and right eye movements [LOC and ROC], and chin muscle tone [EMG]) from the last eight minutes of a thirty-minute REM period are shown. (With permission from Gackenbach and LaBerge, eds., Conscious Mind, Sleeping Brain, *New York: Plenum Publications, 1988, pp. 135-152.)*

"The experiment was set up so that the subject agreed to give a very specific or improbable sequence of eye movements such as left, right, left, right. Figure 4.1 shows a record of one such subject from the last eight minutes of a thirty-minute REM period. The top line is the EEG, the second line is left eye movement, the third line is right eye movement, and the bottom line is muscle tone. The muscle tone line is flat because we are in a REM period. Upon awakening into lucidity the subject made five previously agreed-upon eye movement signals. The first set signal 'left, right, left, right,' is circled (1). This was agreed upon to indicate the onset of lucidity. After some 90 seconds, the subject realized he was still dreaming and signaled again with three pairs of eye movements. He then remembered that the signal was only supposed to be a sequence of two, so he correctly resignaled with two pairs (4). Finally, on awakening 100 seconds later, he signaled appropriately four times 'left, right' in sequence (5). The muscle tone increases when he's awake."

His Holiness was obviously delighted with the experiment, and proceeded to question it in detail. "Was the person who gave these signals able to control his dream and how old is this person?" Jayne replied that the dreamer could control his dreams; his name was Daryl and he was in his mid-to-late twenties at the time of the experiment. "When Daryl was giving the signal, and recognizing the dream as a dream, if you were to speak to him, would he be able to hear you?" Jayne explained that in REM it was very difficult to achieve such incorporations, but there was a case on record in which it had been done. His Holiness added, "In Tibetan dream yoga practice, one method used is to instruct the sleeping person softly, 'You are now dreaming,' once you have an indication that they are dreaming."

He moved on to other details: "During the REM state, the muscles are paralyzed. In that case, how do you explain the phenomenon of emission while dreaming; that is, in dreams in which someone has sexual intercourse and actually has orgasm?" Pete clarified that this was a reflex. The penis muscles involved are not skeletal muscles, and only skeletal muscles are paralyzed during REM.

"Although there is a lot more to be said about these experiments and others that have been done,[13] I prefer to move on to other psychological and social studies, since these electrical studies are not my domain of expertise," said Jayne, changing her transparency. "However, they do seem to put lucidity into the realm of valid phenomena for sleep research, and that is very important."

How Common Is Lucidity?

"In the United States," Jayne continued, "only about fifty-eight percent of people have had a lucid dream once in their lifetime. Maybe twenty-one percent have a lucid dream once or more in a month. In other words, lucidity is still rather rare. However, in another sample of people who had done either Buddhist or transcendental meditation, the average goes up to once or more a week. Here we are not talking about meditators who are specifically practicing dream yoga, but meditators in general."

His Holiness added a reflection: "Maybe this can be seen as an indication that these people have a higher degree of mindfulness. During the dream state there clearly is a form of consciousness in which one may engage in certain types of spiritual practice. For example, one might engage in deity yoga, a Vajrayāna practice, or one might cultivate compassion or insight. But during the dream, if one feels compassion, it is genuine compassion that arises. Experientially, tears can flow from the eyes out of compassion; it really seems to be genuine compassion. There's a doubt though, as to whether this compassion is really significantly different from compassion during the waking state. If you looked at the EEGs for the experience of compassion in the dream state and the waking state, would there be any difference in the patterns?"

"I don't think the experiment has been done, Your Holiness," I ventured to answer. "Remember that these EEG measures are very general. If you look at a person's EEG, you cannot tell if he or she is full of compassion, or completely oblivious. Perhaps if the experiment were done we would not see a great difference between REM sleep and waking in patterns of activity relating to emotional tone."

Traits of Lucid Dreamers

"We were interested in what happens in a regular dream compared to what happens in a lucid dream," Jayne went on. "Are they the same except that you know you're dreaming, or are they different? It depends on who you ask. If you ask the dreamer whether a lucid dream is the same as a nonlucid dream, they claim that lucid dreams are very different: more exciting and vivid. In contrast, if you ask judges to read transcripts of lucid dreams and nonlucid dreams, they feel that there is almost no difference between the two. In statistical analyses, we found that there is more body movement in lucid dreams, and more sound. Together these two facts led us to look at the idea of balance. Body balance seems to be very important for lucid dreaming, not only in the dream state but also while awake. Physical balance is important, as in flying in a dream, but also emotional balance: I want to do something in the dream, but I have to remember that I'm dreaming, so I'm juggling two thoughts. We speculated that it might be linked to the vestibular system of body balance which is tied to the production of eye movements in sleep. Interestingly, we found that there were fewer dream characters in lucid dreams than in nonlucid dreams. All this leads us to ask whether there are psychological, cognitive predispositions to lucid dreaming. It turns out that there are, notably in the realm of spatial skills such as body balance."

His Holiness pointed out that meditators, who have a higher than average degree of mindfulness, also seem to be more susceptible to lucid dreaming experiences: "Perhaps meditators have special skills, since they reflect a lot on their bodily energies and on their bodily and mental states. Perhaps this makes them more in tune with their bodily states. Would you also expect that people's ability to learn to dream lucidly is related to their degree of intelligence?"

"There's a little bit of evidence for that, but in general it is less important than a sense of body orientation in space. Some people get totally lost in the woods, or in the streets of an unknown town. Other people know where they are very quickly, not because of what they see, but because they have a sense of bodily direction. People who have that skill naturally are more likely to have lucid dreams. Body orientation seems also to increase with meditation, by the way. Another factor is complex spatial skills, such as the ability to solve mazes. Lucid dreamers can do those things well. Finally, they have more waking imagery on the verge of sleep, and they also daydream more.

"Personality traits are a third dimension, much less influential than spatial skills. Lucid dreamers are often people who lean towards an androgynous temperament, and those who are willing to take internal risks, such as trying a new drug or shamanic drumming. They're very oriented to an awareness of themselves."

Inducing Lucid Dreaming

"How can one increase lucidity? One can do things before going to sleep, such as cultivating the intention for lucidity. Meditation is another possibility; some people will wake up about three-quarters of the way through their sleep cycle, at about four o'clock in the morning, meditate, and go back to sleep. That seems to help."

"There are a lot of people who mix up their sleep with meditation, but not quite intentionally," His Holiness quipped, and we all laughed with him.

Jayne continued, "Incidentally, women report more lucid dreams than men, but that's because women remember more dreams than men. If you remember your dreams, then you're more likely to remember a lucid dream. About a third of lucid dreams begin as nightmares. Another third begin by recognizing bizarre inconsistencies, such as, 'That's weird—my mother doesn't have a purple face. This must be a dream.' It turns out that naps are really good times to have lucid dreams, too."

"That seems quite likely," said His Holiness, "because that type of sleep state is rather subtle. The person is asleep, but not in deep sleep—mindfulness is stronger. It's much easier to apprehend dreams, too, if you sit up, rather than lie down while sleeping. You should sleep, if you can, with the spine erect."

At this point, as we had agreed, Bob Livingstone, an observer at this conference, took the floor to present a gift to His Holiness from Stephen LaBerge, the researcher who had conducted the famous signaling experiments in lucid dreaming. It was a compact device to help people develop lucid dreaming and remember their dreams better. Bob explained it as a training instrument: a mask worn on the face while sleeping, with a small signaling light so the machine can communicate with the sleeper. The mask is attached to a small computer. Sensors distinguish when the user is in REM sleep, and the computer then gives a gentle signal. After a bit of practice, a user can recognize that the machine has just signaled the beginning of REM sleep, and a lucid dream is likely. The user can then make a conscious effort to be aware of the dream and remember it. The device also keeps track of the number of times one has REM sleep during each night, and the totals for a week or a month. "Dr. LaBerge also wants you to know that the instrument is in a process of evolution," Bob added. "There is an opportunity now for people in the Buddhist tradition to give advice on further development."

His Holiness was obviously interested. "This would be very good for practice while sleeping and dreaming. Sometimes, if you have a strong dream at night, when you wake up it affects your emotional state in the morning. With this we could cultivate wholesome states of mind while dreaming, and that would be of benefit."

Lucidity and Witnessing

After some handling of LaBerge's device, the session moved back to Jayne. "I would like to move now into a less common ground, but one I think interesting for our dialogue here: witnessing dreaming. In contrast to lucid dreaming, *witnessing dreaming* is an experience of quiet, peaceful inner awareness or wakefulness, completely separate from the dream. In witnessing dreaming it's said that the person can manipulate the dream, but simply has no wish to do so. Whatever the content of the dream is, one feels an inner tranquillity of awareness that's removed from the dream. Sometimes one may even get caught up in the dream, but the inner awareness of peace remains.

"Finally, I want to introduce a third state called *witnessing deep sleep*. This is described as dreamless sleep, very likely a non-REM condition, in which you experience a quiet, peaceful inner state of awareness or wakefulness—a feeling of infinite expansion and bliss, and nothing else. Then, one becomes aware of one's own existence as an individual, which may lead to awakening.

"Let me illustrate these states first with the case of a mathematics professor who had practiced transcendental meditation for twenty years." At the prompting of one of the translators, Jayne gave a brief picture of this kind of practice. "Transcendental Meditation is quite different from basic Buddhist meditation: it is absorptive, done with eyes closed and repeating a mantra. It comes from a Hindu lineage and was introduced in the West recently." His Holiness consulted with his Geshe colleagues before returning his attention to Jayne. "This is what happened to this one subject over years of steady practice. In the beginning, this person talked about lucid dreams he had in which the actor was dominant. Here the role of the observer is to recognize that the self is dreaming, but despite this recognition, the feeling still exists that the dream is out there and the self is in here. When you're in the dream, the dream still feels real.

"As you become more familiar with lucidity it may occur to you that you can manipulate, change, or control the dream. In a second stage it occurred to this dreamer that what is 'out there' is actually in some sense 'in here.' The dreamer may actively engage the dream events or control and manipulate them.

"In a third stage his dreams became short. He described them as being like thoughts that arose, which he took note of and then let go. The action of the dream did not grip him or cause him to identify with it as it did in the first stage, where the focus was more on active participation.

"In a fourth stage he discovered that an inner wakefulness dominated. He was not absorbed in the dreams but in witnessing. The dreams were more abstract and had no sensory aspects to them: no mental images, no emotional feelings, no sense of body or space. There was a quality of unboundedness. I quote him: 'One experiences oneself to be a part of a tremendous composite of relationships. These are not social or conceptual or intellectual relationships, only a web of relationships. I am aware of the relationship between entities without the entities being there. There is a sense of motion, yet there are no relative things to gauge the motion by; it's just expansiveness. There are no objects to measure it. The expansiveness is one of light, like the light of awareness, visual but not visual, more like light in an ocean, an intimate experience of the light.'

"Other subjects report needing to let go of lucidity, and move through nonlucidity and nonawareness before developing the witness in sleep. This different sequence may occur if one becomes too attached to lucidity, especially to the active controlling aspect of the self-awareness in sleep. Such an attachment would require letting go of that self-representation in sleep in order to shift to the next state, witnessing. I conducted a study with sixty-six elite Transcendental Meditators. We used these people because these states are so subtle. You couldn't just ask a college student to do this; they wouldn't know what you're talking about. We felt these people would know and remember these states. We received fifty-five lucid dream descriptions, forty-one witnessing dreaming descriptions, and forty-seven witnessing sleep descriptions from the group of sixty-six people, who had been meditating on average for about twenty years. I read through all of the reports and let their own experiences lead my analysis, which is phenomenological and qualitative."

To make sure of the ground we were standing on, I asked for clarification: "We can verify that a lucid dreamer is in the REM state by the signaling in the experiment. How do we know that these people are in witnessing?"

"From their self-report only. This is truly a phenomenological analysis," replied Jayne. "To conclude, let me summarize these tentative observations with a diagram by Fred Travis (fig. 4.2). He suggests that waking, sleeping, and REM dreaming emerge out of a pure consciousness, a silent void. Where each state meets the next there's a little gap, in which Travis postulates that everybody very briefly experiences transcendental consciousness. When we go from sleeping to dreaming, or from dreaming to waking, these little gaps or junction points occur, and so he calls this the *junction point model of mind*."

"This is quite similar to a Buddhist explanation of these little interludes of the clear light of sleep," said His Holiness. "This is precisely the continuity of the very subtle mind. The major occasions are the times of dying, the *bardo*, and then conception. These are junctures, if you like. The subtlest clear light manifests at the time of death, which is one of these junctures. These three occasions of

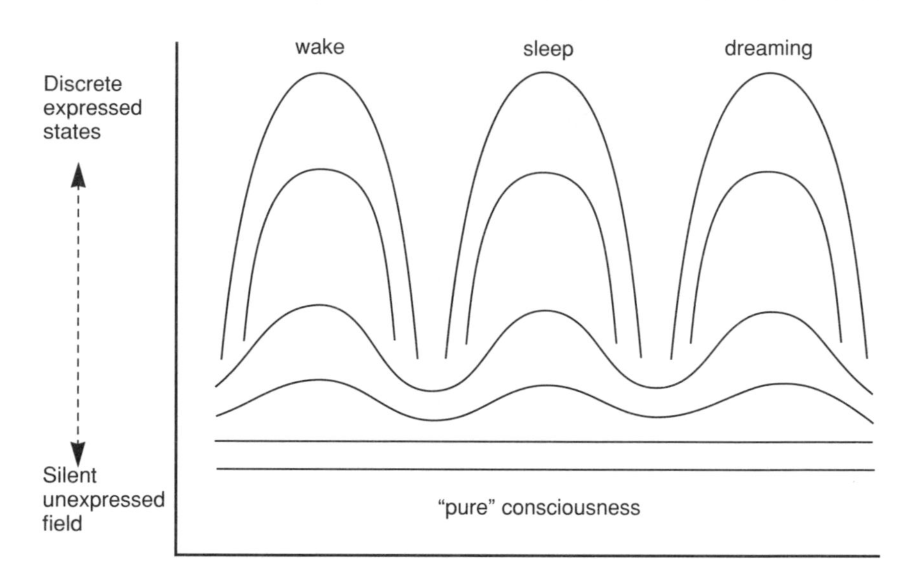

Figure 4.2

F. Travis's "junction point" model for state transitions between three basic forms of consciousness. (Adapted with permission from Travis, "The Junction Point Model," *Dreaming 4 (1994): 72-81.)*

death, *bardo*, and conception are analogous to the states of falling asleep, the dream state, and then waking. There's also a facsimile of the clear light of death in the clear light of sleep. It's not the same as the clear light of death, but it is analogous to it, though less subtle."

Jayne's presentation had concluded and the time for lunch had arrived. It was very apparent that a deepening of the teaching on dream yoga was necessary. It had been scheduled tentatively for the evening, but now the occasion seemed perfect, and His Holiness agreed to open the afternoon gathering with that teaching. It would prove to be a rare treat.

5

Levels of Consciousness and Dream Yoga

As we all settled into our seats at two o'clock sharp, His Holiness began briskly. "Most of you have already heard teachings on dream yoga, but it's likely that this will be fresh for a few of you who have never heard them before."

The Notion of a Self

"I will begin with a discussion of the self. As many of you already know, the basic foundation of the entire Buddhist doctrine is known as the Four Noble Truths. What is the point of recognizing these Four Noble Truths? What is the point of discussing them? It has to do with our basic longing concerning happiness and suffering, and with specific causal relationships. How does suffering arise? How does happiness arise? The central theme of the Four Noble Truths is the issue of *causality* as it pertains to happiness and suffering.

"This explanation focuses specifically on natural causality, instead of invoking some external creator or primal substance that controls the events in life. The Four Noble Truths are often expressed in the form of four statements: recognize the Noble Truth of suffering; abandon the Noble Truth of the source of suffering; accomplish the Noble Truth of cessation of suffering; and cultivate the Noble Truth of the Path. All of this is to be done by the individual who seeks happiness and wishes to avoid suffering.

"In this context, the notion of the self becomes crucial. The person who is experiencing suffering is oneself, and the one who needs to apply the means to dispel suffering is also oneself. And the cause for this is within oneself. When Buddhism first appeared in ancient India, a fundamental distinction between Buddhist versus non-Buddhist views concerned the self. Specifically, the Buddhists refuted the existence of a permanent, unchanging self. Why? Because the very notion of an unchanging self, when applied to the self as an agent and to the self as the experiencer, is very problematic. From the very beginning, there was a great deal of thought and discussion concerning the nature of the self.

"According to non-Buddhist treatises, a self does exist quite separately and autonomously from the aggregates—the psychophysical constituents—of the body and mind. In general, all four philosophical schools within Buddhism agree in denying the existence of a self that has a separate nature from the aggregates. However, these schools have different views concerning how the self exists among the various aggregates of the body and mind. For example, one Buddhist school asserts that the self is the collection of the five psycho-physical aggregates (Skt. *skandhas*). Another school identifies the self with the mind. Within this approach, again there are various views. For example, as I mentioned yesterday, one school asserts that mental consciousness is the self. Then, if you go to the Yogācāra school, you find the assertion that the *foundation consciousness* (Skt. *ālayavijñāna*) is the self.

"And now we move to the Prāsaṅgika Madhyamaka school. According to this school, all of the five aggregates are said to be experienced by the self. Since they are experienced by the self, it becomes problematic to assert that the self is also to be found among those aggregates. It is very problematic if the experienced object and the experiencer turn out to be exactly the same thing. For this reason, the self is not identified as being among the five aggregates. But if you then try to posit a self existing apart from the aggregates, it's nowhere to be found. So that, too, is refuted. The conclusion they draw from this is that the self is designated, or imputed, upon the basis of the five aggregates. For this reason it is said to be merely a name, merely an imputation.

"Nāgārjuna, who is the founder of the Prāsaṅgika Madhyamaka school, says in his *Precious Garland* (Skt. *Ratnāvalī*) that the person is none of the six elements that constitute the person—not the earth element, water element, and so on. Nor is the collection of these elements the person. The person also cannot be found independently of these elements. Just as the person is neither any of the individual elements nor the collection of these elements, similarly each element that constitutes the person can be subjected to the same analysis. They, too, can be found to be merely labels, or designations. Since the person does not exist as a self-subsistent entity that possesses a self-nature or self-identity, the only alternative left is to accept that the person exists nominally or by designation only."

Self and Action

"What is the reason for this very strong emphasis within Buddhism on analyzing the very nature of the self? First of all, the analysis has to do with the self as agent and the self as experiencer. In this sense it's very important. But now let us look to the flow of our experience: feelings of sadness and so forth arise in response to certain experiences. Then certain desires arise in our consciousness. From such desires the motivation to act may arise, and together with this motivation to act comes a sense of self, of 'I.' Together with this sense of 'I,' a stronger sense of grasping onto the 'I' arises; and this may give rise to certain types of mental afflictions, such as attachment and anger. If the sense of 'I' is very strong, then the resultant attachment or anger will likewise be very strong.

"Now one can ask: will the resultant mental states from this grasping onto the 'I' necessarily be of an afflicted nature—for example, attachment and anger—or might they possibly be wholesome? This needs to be examined. In the course of all of this, the issue of the self becomes very central. It becomes imperative to investigate carefully the nature of the self that is involved in these various mental processes. Remember that, from a Buddhist perspective, these mental processes are explained in terms of causality, without positing a self as an agent or experiencer outside the causal chain. This is very important, because our primary concern is seeking happiness and avoiding suffering; and the agent who has this concern is the self. Similarly, when we talk of experience, most of our actions result from our motivations, and all these motivations are ultimately based on a sense of self.

"Effective action has much to do with motivation. Sometimes actions may occur spontaneously without prior motivation, but most such actions are ethically neutral, resulting in neither pleasure nor pain. There's no *absolute* basis for distinguishing between positive and negative action. But by nature we want happiness, so we consider happiness to be positive. Accordingly, those actions and motivations that bring happiness are regarded as positive, while those that eventually create pain are considered negative.

"Now we can ask: is it a bad thing to have a sense of self? The answer, first of all, is that it makes no difference whether or not you want to have a sense of self—it is a given. This sense of self may lead to suffering, or it can lead to happiness. There are also different senses of self. For example, there is a sense of self in which one grasps onto the self as being truly, inherently, existent. Another sense of self has no grasping onto the self as being inherently existent.

"I am persuaded that a strong feeling of 'I' creates trouble. However, that same mental feeling is also sometimes very useful and necessary. For example, a strong feeling of 'I' or 'mine' creates trouble when we make a demarcation between attachment for my friend, and hatred towards my enemy. On the other hand, a strong feeling of 'I' can also create the willpower to succeed or change regardless of obstacles. That is most important. It's not an easy task to develop the mind, and for any difficult task we need determination and effort; and tireless effort comes with a strong will. So in order to develop self-confidence and a strong will, this strong feeling of 'I' is necessary.

"Within a very strong sense of self, what element creates trouble for ourselves? What, exactly, is afflictive? This requires very precise investigation. Through analysis we come up with a threefold categorization of different modes of apprehending the self: (1) apprehending the self as being truly existent; (2) apprehending the self as being not truly existent; and (3) apprehending the self without making any distinction as to whether it is truly existent or not. It's very important to recognize the exact meaning of the phrase 'apprehending as truly existent.' Here, 'truly existent' implies existence by its very own nature."

Motivation for Action Is Mental

"Motivation is a critical factor in the basic aspiration to experience happiness and avoid suffering. What determines motivation? The body may act as a contributing factor, but the chief influence for the formation of motivations comes from the mind.

"To repeat, motivation is the key that determines the nature of our experience, and it is our attitudes and ways of understanding that chiefly influence our motivations. The negative, or afflictive, forces that we are trying to eliminate are also mental in nature. Likewise, the instrument that we are using to eliminate, or at least weaken, these afflictive forces is also mental. Certain mental factors are used to eliminate other mental factors. For these reasons, a discussion of the nature of mind and mental factors becomes very important.

"When you speak about the utter elimination of these faults, you're speaking about something that is a very high attainment: liberation or enlightenment, which may be far away. But in terms of our own experience, it is possible to reduce these afflictive elements in the mind by using the mind. This is something we can ascertain with our experience. To take an example, we all start out with ignorance, a mental process. To attenuate that ignorance we engage in studies and acquire new experiences, and in this gradual process ignorance is lessened.

"In order to transform the mind it's important to have a clear understanding of the mind. For example, the Buddhist Vaibhāṣika school asserts that perception is naked, which is to say that there is nothing interceding between perception and its perceived object. Perception is unmediated. The Sautrāntika school and the two Mahāyāna Buddhist schools say there is a kind of image (Skt. *ākāra;* Tib. *rnam pa*) that mediates between perception and its object. This is similar to the idea of sense data mediating between perception and the perceived object."

Here Charles Taylor intervened to clarify a point of ethics that His Holiness had raised when he spoke of how a strong sense of "I" creates affliction. "Is there any way in which the sense of self is actually wholesome? What distinguishes a wholesome sense of self from an unwholesome one?"

His Holiness responded, "It is very important for someone seeking to overcome suffering to be able to distinguish between the two, for that is a crucial factor in determining our experience. To reemphasize a previous point, there is no *absolute* criterion for distinguishing between a wholesome and an unwholesome sense of self. Rather, in lived experience, you may note that when a certain sense of self arises, together with other mental factors and motivations, this eventually leads to suffering. Because of the nature of that result, you can conclude in retrospect that that sense of self was unwholesome. So it's not a matter of an absolute quality within that sense of the self, but rather it is a relational quality, judged in relation to the results it gives.

"Let's set aside just for the moment the distinction between the sense of self being wholesome or unwholesome and address another related factor, namely, whether a sense of self is in accord with reality or not. Generally speaking, a wholesome mind must be in accord with reality. Moreover, if one is to take one's wholesome mind to its ultimate state, to carry it as far as it can possibly go, it must necessarily be in accord with reality. So let's first of all analyze those three types of senses of self that we mentioned previously to see which of them are in accord with reality and which are not.

"Let's examine the first one, the sense of self that apprehends the self as being inherently existent. How does one determine whether or not this mind is in accord with reality? You can check by investigating whether or not the self that is apprehended in that way exists. Simply put, if there does exist an 'I' which is the referent of this sense of 'I' apprehended as being truly existent, then that sense of 'I' would be in accord with reality. But if there is no referent for that sense of 'I'—if that 'I' in fact doesn't exist at all—then that sense of 'I' is invalid. In Buddhism, this is exactly where the discussion of emptiness comes in."

Charles tried to nail this down: "So in Buddhism that sense of 'I' does not exist by itself?"

His Holiness qualified this position: "That's not true of all Buddhists, or of all Buddhist schools. The term *selflessness*, or *identitylessness*, is universally accepted in Buddhism but the meaning of that term varies from one school to another."

"I thought that the two predicates 'truly existent' and 'existent by imputation' were incompatible. Aren't they?" Charles asked. "We were told earlier that the belief that the self exists only by imputation is common to all Buddhist schools."

His Holiness explained, "There are four major schools of Buddhist philosophy, among which we consider the Prāsaṅgika Madhyamaka school to be the most profound. One school of thought identifies the self essentially with consciousness, whereas the Prāsaṅgika school regards the self as something imputed on the basis of the collection of the aggregates, or the mind and body. The Svātantrika Madhyamaka and all the lower Buddhist schools regard the statement that phenomena exist merely as imputations, not by their own nature, as an expression of nihilism."

"And *nihilism* is a pejorative term?"

"Yes. And from the Prāsaṅgika's point of view, all these other schools erroneously hold to various forms of essentialism, or substantialism."

"Then do you maintain that the self is not truly existent?"

"If in fact the self does not truly exist, then apprehending the self as not being truly existent is, of course, in accord with reality."

"Then the third possibility, not distinguishing one from the other, that must be an error, too, or is it not?" Charles asked.

"When one thinks casually, without any strong sense of 'I' at all, 'Maybe I'll go over there,' or 'I'll have some tea,' or 'I'm feeling like this,' in such cases the sense of self does not, by and large, distinguish the self as being either truly existent or not truly existent. But as soon as the sense of self arises more strongly, for example, when thinking, 'Oh, I'm going to lose out!' or 'I must do something,' then, in most cases, the stronger sense of self comes along with a sense of grasping the self as being truly existent."

His Holiness continued: "For a person who has investigated whether or not the self is truly existent, and through this investigation gains some actual experience of the lack of true existence of a self, when for this person a sense of self starts to arise more strongly it would not arise with the sense of apprehending the self as truly existent. Rather, the self would be apprehended without the qualification of being either truly existent or not truly existent. It could also happen for such a person that, although the self appears as if it were truly existent, one knows that it's not. In this situation, the self is apprehended as being like an illusion. It appears in one fashion but one knows it doesn't exist according to that mode of appearance. Hence it's like an illusion."

Dense as it was, this exchange provided a vivid sense of how theories of mind and ethical behavior are not truly separate in Buddhist tradition. Now, His Holiness was ready to continue on the topic of consciousness.

Levels of Consciousness

"Speaking about the body and mind, the five psycho-physical aggregates include the aggregate of consciousness. When you speak of it in that way, it seems like consciousness, or mind, is a thing existing in and of itself. This is a false representation, because there are many degrees of subtlety of consciousness. For example, the gross level of mind and energy exists in dependence upon the gross physical aggregates. As long as the brain is functioning, there is gross consciousness, and as soon as one becomes brain-dead, one has no more consciousness at this gross level. In the absence of a properly functioning brain, gross consciousness will not arise. So far, this Buddhist perspective accords with the neurosciences.

"The point at which these two traditions diverge occurs in the Buddhist assertion of a vital energy center at the heart, in which the very subtle energy-mind is located. Some Tibetan commentaries say that the heart center of vital energy is actually located in the physical organ of the heart. I would say that's not true, but I'm not really sure exactly where it is located!" He chuckled with delight. "However, when contemplatives concentrate very strongly at the level of the heart, strong experiences are felt, so there is some connection. At the same time, nobody can really say exactly where this heart center is located. Furthermore, there are also differences among Buddhist scriptures pertaining to meditation, philosophy, and the like; and Tibetan medical literature presents its own unique theories concerning the subtle channels, the centers, and so forth. Also, among different tantric systems you'll find certain discrepancies and variations."

Types of Causal Connections

"It's quite clear that consciousness depends on the functioning of the brain, so there is a causal relationship between brain function and the arising of gross consciousness. But here is the question I continue to consider: what *type* of causal connection is it? In Buddhism we speak of two types of causes. The first is a *substantial cause*, in which the stuff of the cause actually transforms into the stuff of the effect. The second is a *cooperative condition*, in which one event takes place as a result of a preceding event, but there's no transformation of the former into the latter.

"We identify three criteria for a causal relationship to be present between, let's say, A and B. First of all, because A is present therefore B occurs. This counters the notion that something nonexistent could cause anything at all. So, if B is to be caused by A, A must exist. The second criterion counters the notion of a permanent, unchanging cause. It states: if A is to cause B, A must itself be subject to change; it must be impermanent. Then A gives rise in turn to B, which is also impermanent. In short, the second criterion is that the cause must be of an impermanent nature; it cannot be unchanging and permanent. Moreover, the cause must also be the effect of something else. There is no first cause without a preceding cause. The third criterion is that if there is a causal relationship between A and B, there must be some kind of accord between the cause and the effect.

"Let's apply this to the causal origination of consciousness and its relationship to brain function. What type of causality exists here? We have, experientially, two types of phenomena which seem to be qualitatively distinct: physical and mental phenomena. Physical phenomena seem to have location in space, and they lend themselves to quantitative measurement, as well as having other qualities. Mental phenomena, in contrast, do not evidently have a location in space, nor do they lend themselves to quantitative measurement; for they are of the nature of simple experience. It seems that we're dealing with two very different types of phenomena. In this case, if a physical phenomenon were to act as a substantial cause for a mental phenomenon, there would seem to be a certain lack of accord between the two. How could one transform into the other, when they seem qualitatively so very different? This needs to be answered, and we will return to this later."

Foundation Consciousness

"Now let's return to the issue of the *foundation consciousness*. The foundation (Tib. *kun gzhi*) is a term that frequently appears in Vajrayāna Buddhist literature. This sometimes refers to emptiness, which is an object of the mind, and sometimes refers to a subjective awareness, namely *clear light*. In the latter case, the clear light is called the *foundation*, or literally *foundation of all*, because it is the foundation of both the cycle of existence and of liberation, of *saṃsāra* and *nirvāṇa*. However, unlike the Yogācāra assertion about the foundation consciousness, according to Vajrayāna this need not be ethically neutral; that is, this clear light does not need to be something that is neither wholesome nor unwholesome. Why is that? Because through spiritual practice this clear light will be transformed into the mind of enlightenment.

"We also find a different usage of the term *foundation of all* in the Dzogchen, or Great Perfection, literature, where it is used in two ways. First, it refers to the basis of latent propensities, and secondly to primordial reality. However, I'm not entirely sure about the referent of this term in its second usage. In the first case it refers to a particular state of mind. According to the Nyingma order of Tibetan Buddhism, the mind is divided into two types: foundation consciousness, which is the foundation for latent propensities, and pristine awareness (Tib. *rig pa*). Experientially, the foundation consciousness is prior to the experience of pristine awareness. What these two kinds of awareness have in common is that appearances arise to both of them; but—unlike ordinary states of mind—they do not follow after, or engage with, appearances. However, the foundation consciousness differs from pristine awareness in that the former includes a certain degree of delusion.

"Pristine awareness and foundation consciousness both share a common quality in that they do not follow after objects. But it is very important to recognize the distinction between the two. Otherwise, one may well misconstrue the nature of Dzogchen practice, thinking that all you do is sit passively without reacting to whatever appears to your mind. It is a misconception that Dzogchen, or the experience of pristine awareness, means just hovering right in the present, without following after the object. To clear up that misconception, we make this distinction: in the foundation consciousness there is still an element of unclarity or delusion in this passive awareness. Whereas when pristine awareness arises, it is extremely vivid, luminous, and liberating. So there's a radical distinction in the quality of the awareness of these two states, but unless you have experienced the nature of pristine awareness, you could confuse the two.

"A person who is training in this practice experiences these states in sequence. As you're sitting passively, not engaging with the object, first there arises the foundation consciousness. Then following that, pristine awareness arises, which qualitatively is very different. Once you have become well trained in the experience of pristine awareness, you would not have to experience the deluded foundation consciousness first, before the luminous pristine awareness. More likely, you could slip immediately into the nondeluded pristine awareness.

This is a very important point.

"There are three types of pristine awareness. *Basic pristine awareness* (Tib. *gzhi'i rig pa*) acts as the basis for all of *saṃsāra* and *nirvāṇa*, and is identical to the subtle clear light. This is the pristine awareness one experiences at the time of death, but not during the ordinary waking state. It is from this awareness that the foundation consciousness arises. Then, through meditative practice, after the experience of foundation consciousness you can experience a second kind of pristine awareness, namely *effulgent awareness* (Tib. *rtsal gyi rig pa*). The third kind of pristine awareness is called *natural pristine awareness* (Tib. *rang bzhin gyi rig pa*). Where does this natural pristine awareness come in? As a result of meditative practice it is possible to gain direct experience of the subtle clear light, and the subtle clear light so experienced is said to be the natural clear light, as distinguished from the basic clear light. The basic clear light can be experienced only at the time of death."

Continuity of Levels

"Finally, let's pick up one dangling subject, namely the origins of consciousness itself. What is the substantial cause of the first moment of awareness following the conception of a human fetus? In Buddhism there are two views on this, the Sūtrayāna and the Vajrayāna. The Sūtrayāna view generally states that there must be a continuum of consciousness: consciousness gives rise to consciousness. There must be an accord between the cause and effect if one is to transform into the other, and for this reason there needs to be a preceding continuum of consciousness that gives rise to the first moment of consciousness following conception. That's a general philosophical theme in the sūtra context. In addition to the prior continuum of consciousness acting as the substantial cause for later consciousness, latent propensities can also transform into consciousness; so there are two types of substantial causes for the origins of consciousness.

"In the Vajrayāna context you find a more precise discussion of this in terms of *very subtle mind*, also called *primordial consciousness*, or *primordial clear light.* This is said to be the substantial cause of all forms of consciousness. The continuum of the very subtle energy-mind is the foundation of all of *saṃsāra* and *nirvāṇa,* a quality that the Yogācāra school attributes to the foundation consciousness. They have that in common, but there are a lot of qualities that the Yogācāra school attributes to the foundation consciousness that are not attributed to the very subtle mind as asserted in the Vajrayāna. This continuum of very subtle mind is not the foundation consciousness as it is asserted in the Yogācāra, not even conventionally. However, because the continuum of very subtle mind as asserted in the Vajrayāna acts as the foundation for all of *saṃsāra* and *nirvāṇa,* we can call it the *foundation of all.*

"Why do the Yogācāras affirm the existence of the foundation consciousness? The rationale is that they are searching for something that is the self. For reasons of logical argumentation they are compelled to make that assertion. But that is not at all how the assertion of the very subtle mind is made in the Vajrayāna. Vajrayāna does not assert the existence of the very subtle mind as a result of trying to find something that truly is the self."

Attempting to link this account with the idea of the continuum of consciousness, I asked whether the continuum of consciousness is the same thing as the foundation consciousness. His Holiness confirmed that in the Dzogchen context the ever present subtle clear light, known also as natural pristine awareness, or Dharmakāya, is in fact the same as the continuum of consciousness.

Mental Factors and Sleep

His Holiness continued: "We find within Buddhism very precise and elaborate discussions of the nature of the mind. For example, distinctions are made between a mind that knows its object and one that does not know its object. And distinctions are made, for example, between valid and invalid cognition; that is to say, cognition that properly apprehends its object and one that does not. Further distinctions are made between mind and mental functions, and between conceptual and nonconceptual awareness.

"Various classifications are made, but the reason for making these elaborate theories is not simply to gain a precise understanding of the nature of the mind. Rather, it stems from the primary issue of determining how to dispel the afflictive factors of the mind and cultivating those factors which give rise to happiness. These theories of the mind try to accomplish this. The text *A Compendium of Knowledge* (Skt. *Abhidharmasamuccaya*), Ārya Asaṅga draws a distinction between mind and mental factors, and fifty-one mental factors are classified. Among those fifty-one factors there are four variable mental factors, and one of these is sleep.[14] A common characteristic of the four variable mental factors is that they may be wholesome or unwholesome depending on other factors such as motivation.

"In addition to practicing during the waking state, if you can also use your consciousness during sleep for wholesome purposes, then the power of your spiritual practice will be all the greater. Otherwise at least a few hours each night will be just a waste. So if you can transform your sleep into something virtuous, this is useful. The Sūtrayāna method is to try as you go to sleep to develop a wholesome mental state, such as compassion, or the realization of impermanence or emptiness.

"If you can cultivate such wholesome mental states prior to sleep and allow them to continue right into sleep without getting distracted, then sleep itself becomes wholesome. The Sūtrayāna teaches ways of transforming sleep so that it becomes wholesome, but it does not seem to include techniques designed specifically to alter the dreams so that they become wholesome.

"There are also references to the use of certain signs in dreams to judge the level of realization of practitioners. This has to do with Pete's question yesterday about recognizing prophetic dreams. If something like this happens just once, it is not regarded as significant, but if such dreams occur very persistently that would be noteworthy. One needs to examine whether there are other influencing factors to be taken into account."

Clear Light, Subtle Self

"Now we move to Vajrayāna and the four classes of tantra. Among the three lower classes of tantra, although there is much discussion concerning good dreams and bad dreams, good signs and bad signs, there is no discussion about the actual utilization of dreams in practice. However, those same three lower tantras include ways of bringing greater clarity to the dream state through meditating on one's chosen tantric deity (Skt. *iṣṭadevatā*; Tib. *yidam*).

"The Highest Yoga Tantra, which is the fourth and most profound of the four classes of tantra, speaks of the basic nature of reality. In addition to the nature of the Path and the culmination of the Path, or buddhahood, this level of tantra discusses both the mind and the body in terms of three progressively more subtle states or levels: the gross, the subtle, and the very subtle states. In this context, one can also speak of gross and subtle levels of 'I' or the self. Would it therefore follow that there are simultaneously two different selves, a gross self and a subtle self?

"The answer is no. As long as the gross body and mind are functioning, the gross self is designated on the basis of the gross body and mind, and on their behavior. During that time, therefore, you cannot identify a subtle self. But with the collapse of the gross body and mind, at the point of the clear light of death, the gross mind is totally gone, and the only thing left of this continuum is the very subtle energy-mind. At the time of the clear light of death, there is only the very subtle energy-mind, and upon that basis you can impute the very subtle person or 'I.' At that time there isn't any gross 'I' at all, so the two—the gross self and the very subtle self—do not manifest simultaneously. Therefore you avoid the error of two people existing at the same time.

"To return to a question you asked before, Francisco, the designation of subtle self occurs during a special dream state. This is not just imagination; the subtle self actually departs from the gross body. The subtle self does not manifest in all dreams, only in a special dream in which one has a special dream body that can separate from the gross body. That's one occasion when the subtle body and the subtle self manifest. Another occasion is during the *bardo*, or intermediate period between two lives. In order to dispel the afflictions of the mind and to cultivate wholesome qualities, it's optimal to use both your gross mind and your subtle mind, and the latter can be cultivated through the practice of dream yoga. If it is possible to utilize all levels of the subtle and the very subtle energy-mind, this is worthwhile."

The Cycle of Embodiments

"Nāgārjuna presents another benefit of the practices of sleep yoga and dream yoga: to skillfully use the faculties that we possess as human beings on this earth, given our particular nervous system and physical constitution, which is a composite of six constituents. With this constitution we experience three states: death, the intermediate state, and rebirth. And these three states, which characterize our existence as human beings, seem to have certain similarities with the embodiments of a buddha.

"One embodiment is called the Dharmakāya, which can be described as the state of pure cessation of the proliferation of all phenomena. There are certain points of similarity between the Dharmakāya and death, in which all the gross levels of energy-mind are dissolved into the fundamental clear light. Moreover, at the point of death, all the proliferations of phenomena dissolve into the very nature of the sphere of ultimate reality (Skt. *dharmadhātu*; Tib. *chos kyi dbyings*). This is obviously not a person, but a state.

"The second state we experience is the intermediate state, which is the interval between two lives. It's the link between death and arrival into a new physical body at conception. At the point of death, from within the clear light of death arises a form which consists of subtle energy-mind, free from the gross levels of mind and body. This is analogous to the Sambhogakāya, which is the embodiment of a buddha in its primordial form, arising from the Dharmakāya. Both the Sambhogakāya and the special dream body are considered to be subtle forms, as is the form one takes in the intermediate state.

"Conception takes place with the initial formation of the gross body and energies. Similarly, from within the pure form of the

Sambhogakāya a buddha manifests in multiple gross forms, called Nirmāṇakāya, in accordance with the needs of various sentient beings. This is similar to conception. It is important to distinguish here between conception and emergence from the womb. The meaning here is definitely conception and not emergence from the womb.

"These are the points of similarity between the three states and the embodiments of a buddha. We also possess the faculties that allow us to go through these three stages during our existence as human beings, and Nāgārjuna suggests that we utilize these faculties according to tantric techniques of meditation. In addition to the Mahāyāna practice of meditating on emptiness and compassion, one can utilize the clear light of death for gaining insight into emptiness, thus transforming death into the spiritual path leading to full enlightenment. Just as the clear light of death can be utilized as the path leading to the attainment of Dharmakāya, the intermediate state can be used for achieving Sambhogakāya; and conception can be used for attaining Nirmāṇakāya."

Dream Yoga

"In order to train in the path that would allow us to transform death, the intermediate state, and rebirth, we have to practice on three occasions: during the waking state, during the sleeping state, and during the death process. This entails integrating the self with spiritual training. Now we have three sets of three:

1. Death, intermediate state, and rebirth
2. Dharmakāya, Sambhogakāya, and Nirmāṇakāya
3. Sleeping, dreaming, and waking

In order to achieve the ultimate states of Dharmakāya, Sambhogakāya, and Nirmāṇakāya, one must become acquainted with the three stages of death, intermediate state, and rebirth. In order to become acquainted with these three, one must gain acquaintance with the states of dreamless sleep, dreaming, and waking.

"To gain the proper experience during sleep and the waking

state, I think it is crucial to become familiar, by means of imagination, with the eightfold process of dying, beginning with the waking conscious state and culminating in the clear light of death. This entails a dissolution process, a withdrawal. At each stage of the actual dying process there are internal signs, and to familiarize yourself with these, you imagine them during meditation in your daytime practice. Then in your imagination, abiding at the clear light level of consciousness, you visualize your subtle body departing from your gross body, and you imagine going to different places; then finally you return and the subtle body becomes reabsorbed in your normal form. Once you are experienced at visualizing this during daytime practice, then when you fall asleep an analogous eightfold process occurs naturally and quickly. That's the best method for enabling you to recognize the dreamless sleep state as the dreamless sleep state. But without deeper meditative experience of this in the daytime, it's very difficult to realize this dissolution as you fall asleep.

"In the Highest Yoga Tantra practice there are two stages for any *sādhana* or visualization practice: the stage of *generation* and the stage of *completion*. In the stage of generation, the more basic of these two, this whole eightfold process of dissolution is experienced only by the power of imagination; you just visualize it. But in the second stage of practice, the stage of completion, by means of *prāṇa* yoga, including the *vase meditation*, you bring the vital energies into the central channel, and you actually bring about such a dissolution, not just with imagination, but in terms of reality. You bring about such a dissolution, and at a certain level of this practice the clear light will manifest.

"If you've arrived at that point in your experience and practice, then it's very easy for you to recognize the clear light of sleep when that naturally occurs. And if you have arrived at the point where you can recognize dreamless sleep as dreamless sleep, then it's very easy for you to recognize the dream as the dream.

"This discussion concerns the means of ascertaining sleep as sleep and dream as dream by the power of vital energy. That's one avenue leading to that result. Now, going back to daytime practice, if one has not reached that level of insight or experience through the vital energy practice, then during the daytime you accomplish this by the power of intent, rather than power of vital energy. Intent means you have to strive very diligently, with a lot of determination. In such practice, recognizing dreamless sleep is harder than recognizing the dream as dream.

"Different factors are involved in the ability to recognize the dream as dream. One is diet. Specifically, your diet should be compatible with your own metabolism. For example, in Tibetan medicine, one speaks of the three elements: wind, bile, and phlegm. One or more of these elements are predominant in some people. You should have a diet that helps to maintain balance among these various humors within the body. Moreover, if your sleep is too deep, your dreams will not be very clear. In order to bring about clearer dreams and lighter sleep, you should eat somewhat less. In addition, as you're falling asleep, you direct your awareness up to the forehead. On the other hand, if your sleep is too light, this will also act as an obstacle for gaining success in this practice. In order to deepen your sleep, you should take heavier, oilier food; and as you're falling asleep, you should direct your attention down to the vital energy center at the level of navel or the genitals. If your dreams are not clear, as you're falling asleep you should direct your awareness to the throat center. In this practice, just as in using the device sent by LaBerge (see p. 106), when you begin dreaming it's helpful to have someone say quietly, 'You are dreaming now. Try to recognize the dream as the dream.'

"Once you are able to recognize the clear light of sleep as the clear light of sleep, that recognition can enable you to sustain that state for a longer period. The main purpose of dream yoga in the context of tantric practice is to first recognize the dream state as dream state. Then, in the next stage of the practice you focus your attention on the heart center of your dream body and try to withdraw the vital energy into that center. That leads to an experience of the clear light of sleep, which arises when the dream state ceases.

"The experience of clear light that you have during sleep is not very subtle. As you progress in your practice of dream yoga, the first experience of the clear light occurs as a result of focusing your attention at the heart center of the dream body. Although the clear light state during sleep at the beginning is not very subtle, through practice you'll be able to make it subtler and also prolong its duration. Also, a secondary benefit of this dream body is that you can be a perfect spy."

He laughed in his usual style. Realizing how much time the teaching had taken, and how late it was, he got up, bowed to all present, and left. We slowly gathered our notes and pads, resting in the aura of a knowledge that was both vast and difficult to grasp.

6

Death and Christianity

THE TIME HAD ARRIVED TO LEAVE the gentle land of dreams and face the harsh reality of death, the ultimate frontier. The day would be devoted to defining how death happens as a bodily process. With such a subject, it was more essential than ever to start by setting the appropriate context. I had requested that Charles Taylor again give an overview of Western attitudes towards death. This he did with his characteristic conciseness.

Christianity and the Love of God

As soon as the Dalai Lama was seated, Charles sat next to him and launched in. "I'd like to talk about Western attitudes to death, but I want to start a bit farther back. One can't understand Western attitudes without understanding Christian attitudes, and it's hard to understand Christian attitudes to death without understanding certain very fundamental points about Christianity. So I'll start off making some basic points of comparison and contrast between Buddhism and Christian faith. In both cases we have a picture of the human being imprisoned in a fixated understanding of the self, needing liberation from it. In both cases this liberation takes the form of changing our very understanding of who we are. Our identity must be transformed.

"At that point we find the divergence. For Buddhism it seems that the transformation—the change of the identity of the self-understanding—comes from a long discipline of understanding the nature of the reality, or the unreality, of this identity as first understood. In a sense, one transcends the self. In Christianity, Judaism, and Islam, what brings about the transformation is one's relationship to God—you might say, the friendship of God.

"This whole religious understanding is based on a very common human experience. In intimacy with certain people, we find that the world appears different to us. In the company of certain people, we can be different. For instance, in the company of holy people, or people who are very profound, our own compassion can increase and our anger can diminish. I think we've all had this experience here in these very days in your company; we're not quite exactly the same as we are outside. In a sense the whole religious standpoint of Christianity, Judaism, and Islam is based on understanding this particular human phenomenon on a much greater scale. Our relationship to God is an intimate friendship with a very holy being. Just as an intimate friendship with a very holy person can transform us, so the friendship with God can operate this transformation.

"I will call this the *dialogical principle*: the conception of the human being as transformed by dialogue, or by relation with others. This dialogical understanding is at the core of Christianity, and the love of God operates here both subjectively and objectively. It implies both the love that we have for God and the love that God has for us. This concept of God's love for the world plays a very crucial role. The world is understood to exist only as held in the sense of God's love. The closer our friendship with God, the more we can participate in that love. There is an extraordinary convergence here with Buddhism in that, the more our identity changes in this way, the greater our compassion and love for other creatures."

Death in the Christian Tradition

"Let us now see how this affects attitudes to death. But first, a remark about the different ways of thinking in the two traditions: I've been tremendously impressed in these last days how disciplined, exact, and far-reaching is the Buddhist understanding of the nature of being, mind, and death. There's almost a rigorous science here, grounded ultimately on the experience of people who have gone far in that. In Christianity, the experience of those who have gone deep into the friendship with God has led to some fairly rigorous thinking in certain domains, but in the farther reaches of death and the afterlife, it has produced nothing as disciplined and exact as we find in the Buddhist tradition. It is assumed that we cannot fully know this domain and so we are dealing with guesses, perhaps inspired, and generalities rather than final truth. So, from the beginning, we expect a different type of discourse here from Christianity than from Buddhism.

"In Christianity, death must not separate us from God. The point is to be with God. There are various ways of understanding this in images such as heaven and hell. These images have been very powerful in Western folklore, preaching, and literature. Think of the medieval poet Dante's magnificent fresco of the afterlife. These images beg correction by one very important theological fact: we can't understand our relationship with God if we retain an ordinary, secular conception of time. God is not within time in the sense of one instant following another, where we are always situated at one point and therefore not at another point in the stream. So we use the word *eternity*, and what we mean by God's eternity is something very paradoxical—God's being present in some way to all of existence over all of time.

"It follows then that being with God means rising into this dimension of time. Think of time as two-dimensional; what we live now is one dimension, and God's eternity is another dimension. Imagine we are ants crawling along the floor, aware only of ourselves at that point on the floor. But human beings standing above the ants can relate to each other across the whole room. They can be in contact with places that are far in the ants' future as they crawl in that direction. They can speak, as it were, from the ants' future and from their past. This is a picture of how God is present in time with human beings.

"Christian theology expresses this as the communion of saints, understood through the notion of resurrection. There's a paradox involved here. The Christian view of human beings is definitely finite. We live at a certain time and a certain place, but in this other dimension we can commune across these finite limits. In the gospel story of the resurrection, a man whose life ends at a certain point with the crucifixion then begins living fully in this other dimension, and therefore can reach beyond it to be present to human beings. He is present to his followers in fullness of his existence even after his death. There have been various myths in Christendom concerning the continuation of a disembodied soul in endless time, but these are not really Christian ideas. In Christianity, the fullness of existence is capable of rising to this other dimension and therefore transcending its limits."

Attitudes Toward Death in the West

"I now want to look at how this works out in the various ways death is understood in the West. The other piece of background we need here is what we discussed the first day, the turning against Christianity that has happened to a large degree in the West (see chapter 1). Both have affected our understanding of death.

"The understandings of death that arise from the Christian story are very much social understandings, and they appear as elements of legend and folk myths. In modern Western consciousness in the last few centuries, this plays out as a major preoccupation with the death of the *other*. This may sound strange, but let me explain it by contrast. From roughly the thirteenth century to the eighteenth century, Western Christendom was obsessed with the death of 'myself' and with the issue of salvation or damnation. I think this religious phenomenon is one of the most important sources of modern Western individualism, although this is a controversial thesis. There are certain important enclaves of Western Christendom where people are still very preoccupied with this, but since 1800 there has been a shift. The great preoccupation with death now is not so much the death of 'myself', but the loss of loved ones. There's a great historian of Western views of death who's talked about the shift from the death of 'myself' to the death of 'thou,' of the other.[15]

"The loss of a loved one is deeply interlinked with the entire dialogical nature of this culture and civilization. One becomes who one is in close intimacy with someone, and then the bonds are broken. In a way, this flows from the whole culture of Christendom, but there's another side to the story: the rebellion *against* Christianity in the West, which itself was inspired by Christianity. Secularism is a very deeply Christian idea. It starts from the notion that there is something good and right about human beings and we have to see that goodness. It is expressed in the very first chapter of the Old Testament which states at every stage of creation "And God saw that it was good." The friendship of God means seeing the good of creation, particularly of human beings. Part of the power of Western secularism is that it claims to do this more effectively than its religious origin. In the eighteenth century, people who did not believe in God claimed to have a higher view of human beings. They thought of them as perfectly all right just as they are, whereas a spiritual Christian view sees people as twisted or distorted in a certain sense, not understanding themselves."

Secular Attitudes Toward Death

"However, secularism separates two things from the Christian root. It cuts off any deeper understanding of loss, or pain, or evil, and it also tends to negate dialogicality. It leads to a picture in which human beings are free, standing entirely alone. This has led to a very strange and uncomfortable pathology around attitudes to death in the West. It has become much more difficult to face the full reality of death. Many people today apprehend death the way our Victorian ancestors felt towards sexuality. What was the cause of Victorian prudishness? They viewed themselves as perfectly capable of living their ethic effortlessly, and so the disturbances of lust were extremely unsettling, not because they had to fight them but because they didn't want to admit that they existed.

"Something analogous has happened with death in secular society. Our picture of the perfection of nature, or happiness, has been purged of all darkness and evil. Youth, health, and strength become almost a cult. Advertisements show beautiful young people cavorting along beaches in sunshine, and there is no evil, or death, or ill health. Death and loss are troubling to admit.

"A number of years ago I was involved in the beginnings of the hospice movement. Elisabeth Kübler-Ross had discovered the extraordinary attitude of doctors in hospitals who tended without realizing it to turn away from hopeless patients and redirect their attention towards the people they could help. Those who were dying were left relatively unattended. If you asked the doctors whether they should perhaps pay attention to these people who might need to talk about their condition, they would all respond, 'Oh no, they don't want to talk about it.' But experiments proved that those who were dying did indeed want to talk about it. The doctors were projecting their own unease with death onto the dying. A whole movement started in which special wards were set up and volunteers were mobilized in an attempt to live through death with those people. This movement is the story of a recovery of an understanding of the meaning of death, a recovery of a capacity to face death. We will return to this tomorrow with Joan Halifax's presentation.

"The interesting thing is that a culture had screened out death. This links up with certain attitudes in modern Western science as well. Science can be the spiritual ally of the secularist position that attempts to make the world entirely good, without allowing for evil or loss, because it offers itself as the instrument whereby you can fix things up and make them good. I think we're all infected with this scientific attitude. We find it in our belief that we can control things."

The stage was set for Pete Engel to take the hot seat and speak as a physician and scientist.

7

What Is Bodily Death?

WITH HIS WHITE BEARD and casual dress, Pete Engel bears himself well, a master of his own subject as befits his role as professor of neurology and neurobiology at the UCLA School of Medicine and director of the Seizure Disorder Center. Yet his speech reflected a genuine modesty and openness, creating a relaxed and pleasant atmosphere.

"As a scientist, what I have to say is, in all humility, extremely simple compared to what we heard yesterday about the Tibetan Buddhist concepts of death and consciousness. Our Western medical view of death is like turning off a light switch: that's the end.

"Charles Taylor's introduction to Western attitudes toward death was wonderful, because it really puts my position into perspective. I'm a neurologist, a physician specializing in disorders of the nervous system, particularly the brain, and I deal with patients who are dying. Medical science seems more concerned with *preventing* death as something evil than with improving the quality of life. I'm also a neuroscientist, so I look at this from a very detached, hard, scientific point of view, which just can't compare with what we heard yesterday or even with what Charles said this morning.

"Personally, as a small child I was so afraid of death that I couldn't even think about it. Today I still have a great deal of difficulty thinking seriously about the fact that I will die, and accepting the scientific view that when I die, that's the end. So it's a tremendous privilege for me to be able to discuss this issue with you and learn that maybe there's more than I thought."

The Western Medical Definition of Death

"I want to cover medical views on death, coma, and consciousness, but also to cover ethical considerations. I believe that there are extremely important ethical issues in what we, as physicians, do to prevent death and prolong life in unconsciousness.

"The physical requirement for life, as Western scientists think of it, is a body consisting of multiple organs, each of which consists of tissues. Some organs are very complicated and have many tissues; others have only one. Each tissue is made up of many cells, which are really the essence of life. Cells can die, an organ can fail, and ultimately the whole body fails because the cells are not sustained. To be sustained the cells need, very simply, an energy source and waste removal. It's like a fire: you have to feed it with fuel. If you cover the fire so that the carbon dioxide it generates can't escape, the fire gets smothered and goes out. If the ashes pile up, the fire will also go out. Like a fire, we need oxygen, which the lungs provide. We need nutrition: we eat food which the gastrointestinal system converts into nutrients such as glucose that cells in the body can use. The lungs remove the waste carbon dioxide. The kidneys remove toxins that build and the liver breaks them down into harmless chemicals. The heart and blood vessels are the delivery system. All of these things are necessary for life.

"Because we're so afraid of death, modern medical science has invested so much effort, money, and resources into preventing death that virtually all failures of the systems necessary for life can now be overcome. If someone's lungs don't work, we can use artificial respiration. If the gastrointestinal system fails, we can put food into the veins. If the liver or kidney fails, we can replace them with machines or transplanted organs.

"Paradoxically, this creates a problem in defining death. Who do we take these organs from? We have to find somebody who's already dead. To find a heart for transplant, ideally it should come from somebody whose body is still alive. So how do we define death? One approach is to define death as a state where *the brain is dead, although the body may still be alive for a while.* We identify individuals who have had accidents or other catastrophic problems that cause the brain to die without affecting other organs. We then harvest their organs. We give their heart to one person, their liver to another, their kidneys and corneas to still others. It's wonderful to see how so many people can benefit because one person has died. As a physician, it helps to be able to discuss this—for example to show parents how their child's death can be of such great benefit to so many people.

"So how does death happen? The organ systems can fail because of trauma or disease. When they fail, for whatever reason, we either lose an energy source or we build up toxins, eventually leading to death. Infection and other external causes may also kill the cells and prevent the energy from circulating appropriately. In Western science, life is based on this energy.

"But there is also a control mechanism for these systems—the brain. In some cases, loss of the control mechanism can also cause death. This is an interesting phenomenon because we think of the nervous system in two parts, voluntary and autonomic. The autonomic nervous system takes care of the vegetative functions of the body: the heartbeat, the stomach making the juices that digest the food, and to a certain extent, breathing. In ordinary individuals the brain has no voluntary control over autonomic functions. However, the breathing centers are located in the brain stem; if the brain dies, the breathing centers also die. The individual stops breathing and the whole body dies. With artificial respiration, the body can stay alive on a respirator for long periods of time."

His Holiness remarked that it was stated earlier that if one can't breathe, the brain dies. Now the order was reversed: if the brain stem is dysfunctional, then you can't breathe. Pete acknowledged that both are true.

"Metaphorically, we can say that the physical structure of flesh, bones, and skin is the earth. The delivery system is like water. The energy system that supports life is fire."

The Dalai Lama suggested that motility is more commonly associated with the inner wind element, rather than fire. In the Tibetan tradition, fire is more closely associated with digestion. Pete pointed out that, if fire represented digestion, it was related to the energy source. His Holiness continued, "Yes, but the term *wind* refers to motility, or movement, of any kind, not only voluntary movement. For example, even after death there's still movement within the body as it decomposes. That very movement of the cells decomposing indicates the wind element. By definition, if there is movement, that is an indication of the presence of a wind element." He paused for a moment, searching for a reference: "The Tibetan scholar Taktsang Lotsawa mentions in one of his texts on the Kālacakra Tantra that a certain type of wind, or vital energy, exists even in a corpse. He writes this in response to the very commonly known point that in the dying process the vital energies all converge at the heart. So there might be a disparity here between the Guhyasamāja system and the Kālacakra system, both included in Highest Yoga Tantra."

The physician in Pete was getting interested: "Can the brain stay alive and the body be dead, if the body provides all the support systems for the brain? That could only happen if there were complete artificial circulation, which is still science fiction. I had wondered whether what you call the clear light state is in fact death of the body but survival of the brain, but if the wind collapses into the heart chakra, then the brain is also dead during the clear light of death." His Holiness nodded his confirmation, giving Pete the occasion to quip: "So the existence of a state with the brain alive and the body dead is extremely unnatural in all systems!"

A Buddhist Definition of Death

When the general laughter subsided, the Dalai Lama continued on the same track: "Misunderstanding may arise by confusing the Buddhist and scientific definitions of death. Within the scientific system you spoke quite validly of the death of the brain and the death of the heart. Different parts of the body can die separately. However, in the Buddhist system, the word *death* is not used in that way. You'd never speak of the death of a particular part of the body, but rather of the death of the entire person. This conforms to the general, consensual usage of the term death. When people say that a certain person died, we don't ask, 'Well, which part died?' The word *death* is an inclusive term referring to a person, rather than a specific term referring to an individual organ. According to Buddhism, the definition of death has to be understood in contrast to the definition of life. *Life* is defined as the basis for consciousness. As soon as the body is no longer able to support consciousness, there is death. From a Buddhist perspective this is a good working definition in the human context, generally speaking. But if you want to go into greater detail, then you must look beyond human existence, and take account of the formless realm as well as the desire realm and the form realm. The definition of death just given works quite adequately within the desire realm—the realm we're living in—and the form realm, which we have not discussed. But in the formless realm sentient beings do not have gross bodies, so in that context our previous definition of death becomes very problematic."

The Buddhist tradition holds the view that sentient beings appear in forms unfamiliar to us in this material plane or desire realm. Sentient beings exist in a total of six realms: higher beings of pleasure, jealous gods, humans, animals, hungry ghosts, and hell realm beings. From a Western standpoint the definition given by His Holiness also raises questions even for the more familiar animal realms, since many would doubt that consciousness can be predicated for rats and butterflies, although there are some dissident voices.[16] Also, some modern Tibetan scholars have given a more figurative interpretation of these various life forms.[17]

Interlude: A Conversation on Body Transplants

Seeing that tea was waiting at the door, Pete said: "I just wanted to make one more point before we stop. Because the dominant Western scientific view equates the mind with the brain, and the person with the mind, the objective of modern medicine is to keep the brain alive, sometimes at the expense of other organ systems. On the other hand, when the brain dies, we die."

"It's not likely that there will ever be brain transplants?" His Holiness was returning here to the fascinating topic of brain versus body transplant that was raised in the first Mind and Life Conference,[18] and his question set off the following rapid exchange:

Pete Engel: It's an interesting paradox, because that would be a body transplant.

Dalai Lama: If the brain were transplanted into a new body, would that new body become the body of the former brain donor?

Pete Engel: That's right. The person goes with the brain, so a brain transplant is a person transplant.

Dalai Lama: If this is the case, then the person whose body received a new brain would actually not be saved?

Pete Engel: Right, the body is the donor. We speak of the person who gives the heart as being a donor, so in this case the entire body would be the donor.

Dalai Lama: But surely the transplant is constructive; you in fact create a whole different new person.

Pete Engel: If your characteristics as a person are the way you stand, the way you gesticulate, the way you speak, then according to Western science, if your brain was transplanted into my body, my body would take on the characteristics of yours in those aspects that are governed by brain function.

Dalai Lama: The question partly concerns just how many characteristics we are willing to say constitute the core of the person. How much transformation would effect a shift in the core of the person?

Pete Engel: You've asked a very difficult question. Computers are becoming so complicated now that one can say that they think. They can even be said to be creative. This forces us to try to define what is peculiar about the human brain that makes it different from what a computer could conceivably do in the future, to justify that we are human and the computer is not. I don't think any neuroscientist would say that eventually computers will become human, but neither could they give you a good reason why. It becomes a deeply philosophical or even spiritual issue, which is not a satisfactory answer for a pure scientist. I'd like to ask my scientific colleagues here if there is anything to report.

Dalai Lama: If you really identify the person with the brain, we can ask whether a person exists during the fetal formation before the brain. You responded that the general perception is that there is no person at that time. If that is true, that would constitute a justification for abortion. You wouldn't be killing a person, you would simply be removing a part of the mother's body.

Pete Engel: There is a very serious debate in the West concerning at what point in the formation of a fetus it becomes a person. Different schools of thought depend on different religious backgrounds.

Francisco Varela: There is a distinction to make between ending life in a basic form, and killing a mind, a person. Most scientists would agree that the person comes with the brain, and therefore there has to be some kind of brain in the fetus before you have a *person*. But that is not an argument for abortion per se, because you surely have *life* even at the moment of conception.

Pete Engel: We'll come back to the issue in the afternoon. Let me elaborate, though, because you gave us some food for thought about brain transplants. Partial brain transplants are currently being done in Western medicine, although they don't work very well. The entire brain is not transplanted, but in diseases due to destruction of a small part of the brain with a very specific and necessary function, it is now possible to isolate neural cells with the same function from a fetus and inject those cells into the brain. They will grow and make connections and correct the existing deficit. The brain cells have to be in a developing form rather than fully developed in order to grow and connect properly.[19]

Dalai Lama: And the fetus has to die, doesn't it? You can't take it from a living fetus?

Pete Engel: That's right. The brain is extremely complicated, though, and has many different parts with many different functions. We can't just say that the brain is the mind and the person; we're forced to ask what part of the brain is the mind. If we can transplant parts of the brain, how much of the brain can we actually transplant before it becomes a different person? That's an interesting issue!

Interesting indeed, and wide open to ponder for many years. Tea came in, however, and the conversation broke up as usual in smaller groups.

Brain Death

When we reconvened, Pete continued his presentation. "Let me explain *brain death* with a very simple exposition of how the brain dies. If the cause of death is a generalized toxic or metabolic problem, it affects the heart, and the rest of the body also dies. In order to have brain death, you need something that affects the brain but not the rest of the body. If the cause of death is a systemic poison, it will affect the heart and the brain, and none of the organs are good for transplant.

"What ultimately kills the brain is loss of blood flow to the brain; the delivery system becomes destroyed. The brain is in a cranial vault, a box that protects it. It's surrounded by fluid and can remain intact despite a tremendous amount of trauma to the head. But that box is also a prison. If something happens, like a tumor, that takes up space inside the cranium, it pushes against the cerebrum and the brain stem. Where the spinal cord and brain stem join the rest of the brain there's a hole in the bone (called the *tentorial notch)*. This is a possible area of *herniation*, where the brain can push into the space and come out the other side. If you have some expanding structural abnormality in the brain such as a hemorrhage, an abscess, a tumor, or swelling caused by infection or trauma, the brain herniates into the tentorial notch. Not only does this push against the brain stem, which is the critical area for consciousness and for most bodily functions, it also presses against major arteries in the brain stem and stops the blood supply to the entire brain.

"There's a second situation, when the abnormal mass is in the part of the box that the brain stem is in, causing direct pressure on the brain stem and brain death. Within the brain there are holes called *ventricles* filled with fluid. This *cerebrospinal fluid* is made in the ventricles, and it flows through a small tube in the brain stem called the *aqueduct of Sylvius* to get out of the cerebrum. This is the only way that the cerebrospinal fluid can get out, so when you have a mass in the second vault that causes upward herniation it closes off the aqueduct of Sylvius so that the fluid can't get out. The fluid then builds up, causing tentorial herniation and cutting off the blood supply. Either way, the end result is stoppage of the blood supply to the cerebrum, the part of the brain where we believe consciousness lies. The person dies but the rest of the body is alive, which is the state of brain death. These are the best organ donors for people who need a heart or a kidney.

"How does modern medicine define brain death with sufficient certainty to remove an individual's organs or at least terminate their respiration? All these people are on respirators, because the only way this state occurs is if respiration, the one critical factor for body survival that's controlled by the brain, is artificially provided. To keep somebody in this state for weeks is a tremendous emotional drain on their family, and it's very expensive. How long do you keep someone in that state? At what point do you turn off the respirator, let the body die, and say, 'We're finished?'

"There's no argument that if the entire brain, including the brain stem, is dead, then the respirator can be turned off. That requires that there be no brain stem reflexes. Certain reflexes are mediated through the brain stem, and if they're gone we can recognize that the brain stem is dead. Breathing, for instance, is a brain stem reflex. If you take the patient off the respirator and three minutes pass without spontaneous breathing, then that reflex is absent. If the eyes don't move when you put cold water in the ear, then that reflex is absent also. If the eyes move, there is still some function in the brain stem. On the other hand, preserved simple reflexes in the rest of the body that are mediated by the spinal cord have no bearing on the state of the brain. So we can diagnose brain death even if these kinds of reflexes persist because in the West we don't think we live in the spinal cord, we live in the brain.

"The other thing that we consider is the EEG, which indicates whether the cerebral cortex is functioning. If there is EEG activity, then you know the brain is not dead. If the EEG is flat, the brain may be dead, but it's not absolute proof. As Francisco said, it's like placing a microphone in Dharamsala: if you don't hear anything, it doesn't mean there isn't anyone there. So all these things have to be considered together. We also look at the blood flow by injecting dyes into the arteries and using x-rays to see whether blood is flowing to the brain. If there's no blood flow to the brain, that's an absolute indication of brain death. But that's expensive and difficult to do, so usually we rely on the EEG and brain stem reflexes. If the EEG is flat, the brain stem reflexes are gone, and the cause of the coma is known to be irreversible, then the patient is brain-dead and we can turn off the respirator. If we don't know the cause and it could be reversible, such as a drug-induced state, then we can't be certain. We used to repeat all the tests after twenty-four hours to show that the situation was persistent, but in many cases that's no longer necessary."

Brain Correlates of Consciousness

"We believe that consciousness, everything that distinguishes human life, is in the cortex. If the cortex is dead but the brain stem is alive, is the brain dead? That is currently an argument in Western medical science. The opposite can also be true, and is a particularly tragic situation. If somebody has a stroke in the brain stem, for instance, they may be paralyzed from the neck down and unable to breathe. They survive on a respirator and they cannot move. Their brain is still alive, but they have no way of communicating. They can see because the nerves from the eyes go directly into the brain, and they can move their eyes, because the eye muscles are high up. Sometimes they can hear, but they can't talk. It is called *locked-in syndrome* and without careful neurological testing it's very difficult to know that the brain is alive. These people are often treated as though they're in coma, but they can see and hear what's going on."

The mere evocation of such a condition is distressing, and not surprisingly His Holiness wanted to know if a cure was possible.

"It depends on the cause," continued Pete. "Usually the cause is not curable, but it can be. It's usually a stroke, but sometimes the symptoms are caused by the edema, or swelling, that results from the stroke, and then over time the swelling goes away and function returns. The person may not be completely normal but they get movement back. So it is important to recognize a patient who can be saved, and it's equally important to recognize a patient who is awake and understands what's going on and is probably terrified, and not just ignore them as if they were in coma. An EEG for such a person is normal; you can tell when they're awake or asleep. You can communicate with them by setting up a code based on eye movements.

"Let me delve more into the issue of consciousness. In the upper part of the brain stem is an area known as the *ascending reticular activating system*. The lacy network of the reticular system extends all the way through the brain stem and up into the *thalamus*. The thalamus plays a very important part in integrating sensory motor functions, but its most important responsibility is arousal. It's what causes you to be awake. Lesions of the reticular activating system prevent a patient from being awake. In the lower part of the brain stem are the respiratory functions. Because they are separate, lesions can destroy consciousness without destroying respiration. The unconscious individual continues to breathe" (fig. 7.1).

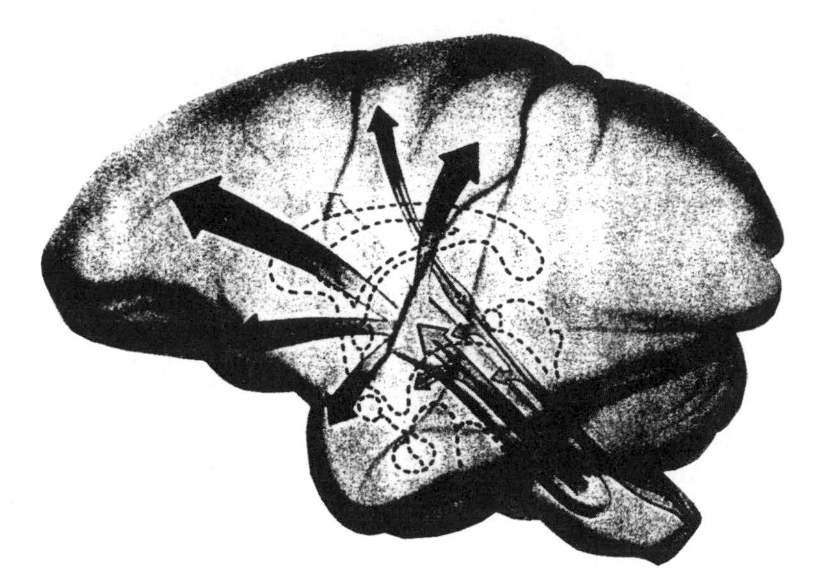

Figure 7.1

Ascending reticular activating system of the monkey's brain. (From Magoun, in Delafresnaye, ed., Brain Mechanisms and Consciousness, *Oxford: Blackwell, 1954.)*

Alterations of Consciousness

"I'm going to talk now about consciousness versus unconsciousness. Unconsciousness includes coma, but there are other types of alterations of consciousness that are worth discussing. The definition of consciousness in medical terms is simply a state of awareness of self and environment. That excludes consciousness in sleep, but to some extent sleep becomes a form of consciousness with dreaming, and certainly with lucid dreaming.

"I want to cover some basic concepts of organization of function in the brain and to show you some new techniques that actually demonstrate these things noninvasively in the living human brain. In figure 7.2 there are a number of fundamental subdivisions of the brain that you have seen before. For example, the frontal lobe, the parietal lobe, the occipital lobe, and the temporal lobe all contain so-called primary cortex, including primary motor cortex, primary somato-sensory, primary visual, and primary auditory cortices.

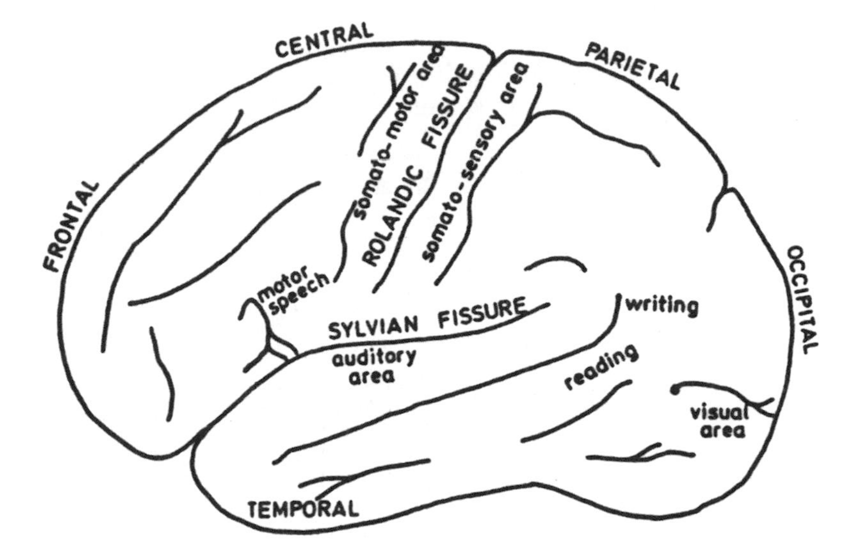

Figure 7.2

Regions of the cortex with specialized anatomical and physiological properties. (Adapted from Cooper et al., EEG Technology, *2d ed., London: Butterworth's, 1974.)*

Then on the left side of the brain in most people is language—motor language and receptive language. These areas are all relatively small. Most of the brain is made up of what we call *association cortex* with functions that are very difficult to define. These parts can often be removed without changing the patient's ability to function or their personality. There is also a tremendous degree of plasticity; if you damage one part of the brain, another part can take over that function. Plasticity is greatest in the very young. As you get older you have less plasticity, but the primary cortices are not plastic.

"Traditionally, there is an argument in neuroscience as to whether higher functions are discretely localized in the brain or distributed. For example, when I think about my mother, does this happen in the hippocampus, a key area responsible for memory, or does it require a dissociated network of two cells here, three cells there, a few more on the other side, to give rise to this mental vision of my mother? The growing consensus is that in fact the whole network is involved. You can remove pieces and in most cases there is enough left to complete the picture. However, small lesions in primary cortex can cause very dramatic neurologic deficits. If they're located elsewhere, they may not cause much problem at all. Altering consciousness requires damaging the entire cortex on both sides, or the reticular activating system in the brain stem.

"Now I want to show you some of the new technologies for brain imaging. What you see in each of the images in figure 7.3 (see color plate between p. 158 and p. 159) is the structure of a normal person's brain with color superimposed on it to provide information about function. The color is quite diffuse, but we can make it more localized by creating special tasks. The color represents information obtained from an imaging technique called positron emission tomography, or PET. It creates an image of function in the brain by injecting a radioactive tracer into the blood and then detecting where it appears in the brain. In this case, the tracer is glucose, so we can see what parts of the brain are using a lot of sugar and what parts are using less. The picture of glucose utilization that the computer generates is then colored so that the highest glucose utilization appears as red, followed by yellow, green, and blue as the lowest utilization.

"The images in figure 7.3 show individuals doing various cognitive tasks. There's some activity in different regions depending on the task. The dark blue spaces are the ventricles filled with cerebrospinal fluid; there's no brain activity there at all. For instance, for the task labeled 'visual' the subject is looking at a pattern in front of him. When the subject opens his eyes, you can see activity in the occipital cortex. This is the visual cortex (shown with arrows). 'Cognitive' tasks required subjects to remember and solve problems, while 'memory' tasks required subjects to listen to a narration and remember as many details as possible. The 'motor' task was simply to sequentially touch the fingers to the opposing thumb. In 'auditory' the stimulus was a combination of verbal and nonverbal material played into both ears. You can see the involvement of the auditory cortex. Almost everybody has a language-dominant hemisphere (usually the left), where the language center resides, and a nondominant hemisphere."

His Holiness's penchant for experimental detail immediately materialized. "Is this specific to one's mother tongue?" Pete explained that the dominant hemisphere would not be activated when listening to a foreign language one didn't understand. The Dalai Lama persisted: "Is there any difference for a person speaking his or her own mother tongue as opposed to a learned foreign language?" Pete again explained, "Language processing is all on the dominant side, but if someone knows two languages they may be located in different areas. A stroke in part of the brain may cause the loss of one language but not the other." "I see," His Holiness answered with a big grin, and then burst into laughter. "I know of a prime minister of India who knew eleven languages, so he's in a much safer situation!"

Pete continued, "This is a fascinating topic because some languages have more right-sided representation than others. Japanese, for instance, is very pictorial and involves more of the right side of the brain, which is essential in visual, spatial organization."

"That would be the case for a person who can read the *kanji* characters, but is it the same for a person who doesn't read?" His Holiness asked.

"This is a very controversial area. Some Japanese linguists believe that certain aspects of the Japanese language mimic sounds of nature, such as bird and insect sounds. The right-sided involvement undoubtedly has to do with the writing, but it's more complicated than that."

His Holiness noted that Tibetan also has many words that mimic sounds. "For example, the word for 'motorcycle' is *bok-bok*," and again we all joined him in a good laugh.

"It may be that the most important function is in an area that we don't see at all. What we can recognize is the number of cells using glucose. If the essential function is very efficient and involves just a few cells, we wouldn't see it. Most people believe that music is perceived on the right side, but that's not always true. For instance, in one experiment people were given a musical task. The task was to listen to a series of tones and then, after a pause, to determine whether a second series of tones is the same or different. Some people do this task on the right side, and some people on the left. This has only been done with a few subjects, but it's interesting that there's a difference in the way people remember the tones. Those that use the right side of their brain remember the tones by humming them over to themselves. Those who use the left side are either trained musicians who visualize notes on a scale, or they use an analytical approach and visualize the notes, for instance as bars of different length. The strategy is different. The left side of the brain seems more involved in analytic processing. To prove this, the researchers took people who had an analytic approach to remembering tones, and did a similar test using timbre instead of tones. *Timbre* is a quality of sound that's easily recognized, such as the difference between a piano and a violin, but it's not easily quantified. You can't visualize it as you can notes on a scale or as lengths of bars. When the people who heard tones on the left side did the same test with timbre, the activity moved to the right side.

"The next thing I want to talk about is mechanisms of altered consciousness. We classify brain injuries by their symptoms as *destructive* or *irritative*. Paralysis, blindness, or deafness are destructive; hallucinations, pain, or epileptic seizures are irritative. Destructive events that cause altered consciousness involve the entire cortex diffusely. But irritative disturbances can also cause altered consciousness or loss of consciousness. Psychogenic disturbances are one example; here the mind rather than the brain alters consciousness.

"Destructive injuries may be acute or chronic. Acute states of altered consciousness can be defined from the least disruptive to the most disruptive. In the *confusional state* a person is disoriented; *delirium* implies even more confusion. In *obtundation* they have difficulty staying awake, but you can wake them and get answers from them. Next is *stupor*, where it's difficult to wake them or get any decent response, but they may respond to painful stimulation. Finally in *coma*, they are totally unresponsive. These terms are not very quantitative. I think they're terrible because they don't really tell you what's going on. To study these situations you really need to be specific about the stimulation used and the response that you get. Coma is a very severe dysfunction where the reticular activating system is in shock, although it may not be damaged directly. In certain types of coma, the damage to the reticular activating system may be irreversible. It's important, however, to know that coma is always a transient state, that people will eventually come out of it: they'll either die or wake up. It may take weeks, sometimes even months, for this shock to be overcome, but if they survive they will eventually wake up. But if they have severe brain damage, they may not be any better off when they wake up than they were in coma. This is called a *persistent vegetative state.*

"The vegetative state occurs when somebody wakes up after several weeks or months of coma, but they have no conscious brain. Their autonomic functions are working but they have no ability to respond to the environment and no awareness of self. They may or may not need a respirator. This is an important ethical issue in medicine. These individuals create a tremendous amount of suffering for their family and require tremendous resources to keep them alive. Right now, if an individual in a vegetative state can breathe on their own, they can live indefinitely and there's nothing that we as physicians can do to stop that.

"*Dementia* is a chronic state in which some structural damage to the brain causes a person to be unable to function in virtually all realms. There are many different forms of dementia, which are usually due to degenerative diseases of the cortex.

"Another type of chronic altered consciousness is *hypersomnia*, increased sleepiness. There are two types. People may be sleepy for reasons that Francisco talked about on Monday (see pp. 36-38). The other is the *neutral state phenomenon* where there are brief periods of sleep, called *microsleeps*. During these periods the individual functions, but is really asleep and will not remember what they did. These people can act out bizarre behaviors. For instance, they may get in a car and drive, and then suddenly find themselves in a strange place, not knowing how they got there.

"There are also psychogenic states. Three occur in *psychoses* or *schizophrenic* situations. In catatonia, the individual is awake but doesn't respond. You can usually put their limbs in certain positions and they'll stay there. It's a psychiatric condition called *waxy flexibility*. It's not due to physical damage in the brain that we can see, but is probably a chemical disturbance of the brain that we haven't yet discovered. *Delusions* are psychotic states where things appear different from what they are. For example, a dog might look like a fearsome lion. *Hallucinations* that occur in psychotic states are mostly auditory rather than visual. An individual may hear voices that tell him to do things. *Hysteria* is a *neurosis*, which is not as profound as psychosis. Individuals show the symptoms of a disease that they really don't have. Coma can be a hysterical symptom: sometimes a patient appears to be in coma but the brain is perfectly normal."

Epilepsies

"The final cause of altered consciousness is epilepsy, the field in which I have specialized for many years. *Epilepsy* consists of states in which the brain cells overreact because of some lesion or other disorder. These states can be subdivided into partial or generalized. *Partial seizures* begin in a part of one hemisphere of the brain, while *generalized seizures* begin on both sides simultaneously.

"Partial seizures are further divided into simple and complex, depending on whether consciousness is altered or not. A *simple partial seizure* may be just twitching of the hand or seeing something that's not there. A *complex partial seizure* results in loss of consciousness. Individuals may fall down, or have strange limb movements, or chewing movements, but they don't have vigorous violent movements. Complex partial seizures are due to abnormalities in a particular part of the brain called the *limbic system* or *temporal lobe*.

"We divide generalized seizures into convulsive and nonconvulsive. A *convulsive seizure* is what most people think of as an epileptic seizure; the individual becomes rigid, falls down, and then shakes. He may bite his tongue or wet his pants. In some *noncon-*

vulsive seizures the patient has just a brief lapse of consciousness, maybe for a few seconds; he might blink his eyes and that's all. Such *absence seizures* may happen several times a day. In *myoclonic seizures*, there's just one or a few quick jerks. There are other seizures in which the person loses muscle tone and collapses.

"These are all involuntary and not at all uncommon. Chronic epilepsy occurs in about one percent of the population. If you live to be eighty, your chances of having at least one of these seizures is one in ten. Epilepsy is a unique phenomenon in the nervous system that has allowed us to understand mechanisms of brain function. Epilepsy has been described in Western medicine for many centuries, although in the past it was incorporated into religious beliefs and seen as possession by the devil. People who have epilepsy are often doubly cursed, first by the disease and then by others' view of them as being possessed or crazy.

"There are two forces in the brain, *excitation* and *inhibition*. Single brain cells secrete a chemical called a *transmitter* which can either excite another cell and make it fire, or inhibit it, and keep it from firing. When the brain stops working, for instance in paralysis or unconsciousness, it may be due to lack of excitation or to increased inhibition. In some cases loss of consciousness is an active process, not a negative process. Epilepsy is a condition of the brain in which cells are abnormally active. If localized they give rise to partial seizures; if they are distributed throughout the brain, they cause generalized seizures. The overall activity can be excitation that makes you do something, or it can be inhibition that makes you unconscious. Synchronization is an important factor. Ordinarily, cells work independently and their independence is necessary for function. If there is too much synchronization, the cells can't do what they're supposed to do. Just as when you play the piano, every finger does something different to make beautiful music. If you did it with your fists, it would make noise: that's an epileptic seizure.

"In figure 7.4, an EEG from a patient with partial epilepsy shows sharp spikes in the right temporal region caused by the abnormally synchronized discharges of neurons in this area. These

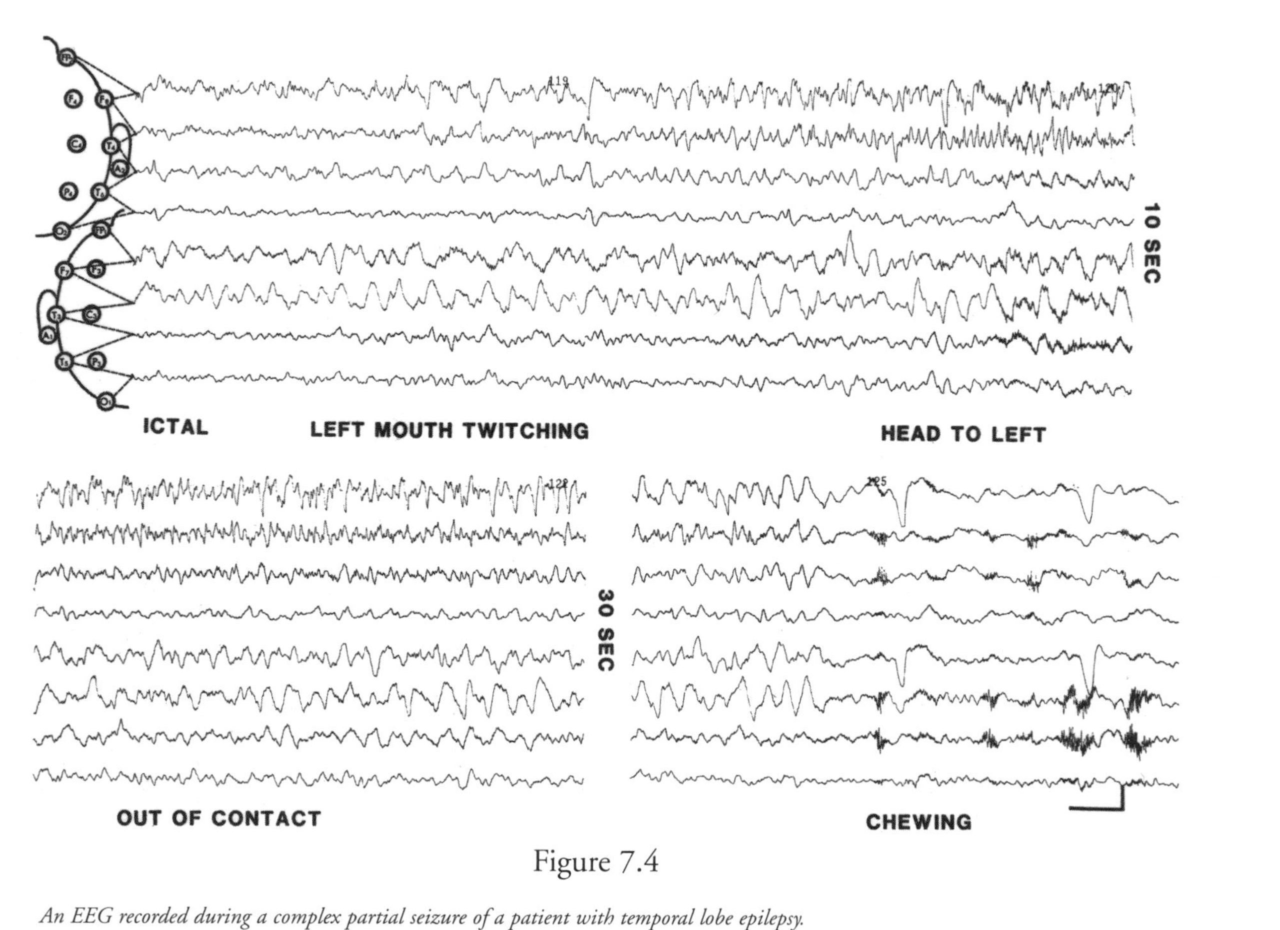

Figure 7.4

An EEG recorded during a complex partial seizure of a patient with temporal lobe epilepsy.

rhythms appear at the beginning of the seizure and gradually spread to involve larger portions of the right hemisphere and then the other side of the brain as well. Initially, the patient had mouth twitching and experienced some strange feelings. He then turned his head to the left. When the epileptic activity spread to the other side of the brain, he lost consciousness for almost a minute. After the seizure, there was suppression of activity so that the brain was not functioning much at all. Although the patient was making chewing movements at this time, he was not responsive. Figure 7.5 (see color plate between p. 158 and p. 159) shows two PET scans. Two sections from the first scan on the left were obtained when the patient was not having a seizure, and show his usual state of relatively decreased glucose metabolism in the right hemisphere, where his seizures begin. On the right are two sections from a scan obtained during the seizure seen in figure 7.4. This is an average of metabolic activity over several minutes, including the seizure seen in figure 7.4 and the period afterwards. The increased glucose metabolism is seen with white as the highest, and indicates where the seizure began and initially spread in the right hemisphere. The decreased activity in the rest of the brain shows the effects of generalized reduced brain function after the seizure was over.

"There are examples of two other kinds of seizures with loss of consciousness. A convulsive generalized seizure is a *tonic clonic seizure* in which the patient becomes rigid and then shakes. This shows that there is increased activity throughout the entire brain. The other type is the *absence seizure*, which is a very brief loss of consciousness. In that situation the EEG looks very different because that seizure is mostly inhibition, not excitation. It lasts a few seconds and the EEG during that time shows high-voltage, sharp and slow waves. The PET scan still shows a tremendous increase in glucose utilization for just these little absence seizures, demonstrating that this inhibition is an active process (fig. 7.4 and fig. 7.5).

"During these states of altered consciousness a patient may see visions if the seizure occurs in the visual system, or hear things if it's in the auditory system. He may do strange things such as shake or wander. In ancient times he would be considered very bizarre. People who had seizures were seen by different cultures in widely different manners.[20] Early Christians believed they were possessed by the devil and burned them at the stake. Other religions have thought that these people were possessed by some good spirit. The early Greeks thought that epilepsy was a blessing and it was called the "sacred disease." It's believed now that the wonderful visions of some great Christian saints like Joan of Arc were in fact epileptic seizures. Mohammed was believed to be an epileptic, as he admitted in his writing."

Epilepsy and Tibetan Medicine

Dr. Engel continued, "I'm curious to know how epileptics are viewed within Tibetan culture. Many years ago I met Dr. Dolma, who told me that epilepsy was not a serious problem among Tibetans like it is in the West. I understand your physician is here today, and I wonder if he can comment on whether many Tibetans have epilepsy, whether you have a treatment for it, and whether you think it is a natural disease process or believe that it has some spiritual significance."

I noticed with some amusement that the specialist in Dr. Engel was not about to miss an opportunity to explore this question. His Holiness turned to Dr. Tenzin Choedrak, his personal physician and one of the most highly respected members of the Tibetan medical tradition. After many years as a practicing physician, Dr. Choedrak had spent years in Chinese prisons before coming to India. Not only was his knowledge remarkable, but I was particularly impressed with the warmth and simplicity of his person. He spoke very calmly through the interpreters: "Although there isn't any extensive discussion of epilepsy and its treatment in Tibetan medical literature, there are references to it. These include descriptions of the symptoms of an epileptic seizure and explanations of the physical dysfunction that leads to such seizures. They speak of three major types of epileptic seizures, related to the different types of metabolism that are detected from the examination of the pulses.

Figure 2.2

Demonstration of EEG recording during the meeting by means of a portable recording setup. Our volunteer, Dr. Simpson, is sitting on the right with an electrode cap, while His Holiness the Dalai Lama looks into the display.

Mind and Life IV participants and translators with His Holiness the Dalai Lama outside of his personal residence in Dharamsala, India.

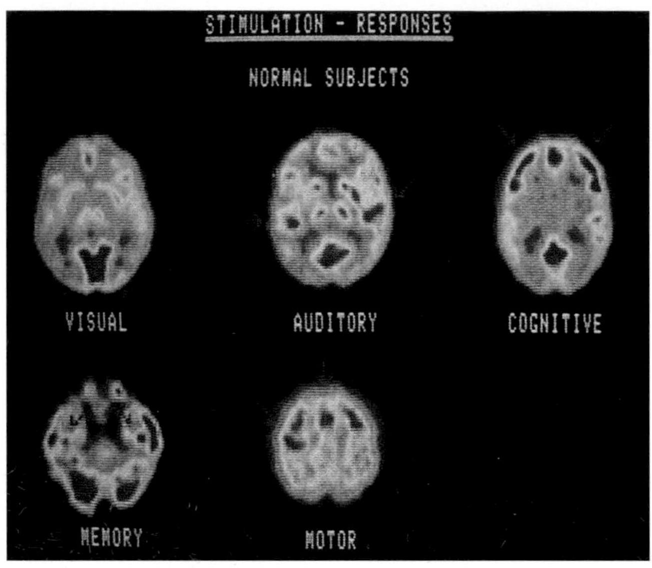

Figure 7.3

Five states of behavioral activation reflected in glucose metabolism of the brain. The highest metabolic rates are shown in red, and the lowest in blue, with intermediate values in yellow and green (Printed with permission from Phelps and Mazziotta, "Positron Emission Tomoraphy: Human Brain Function and Biochemistry," Science 228 [1985]:799. ©1995 American Association for the Advancement of Science).

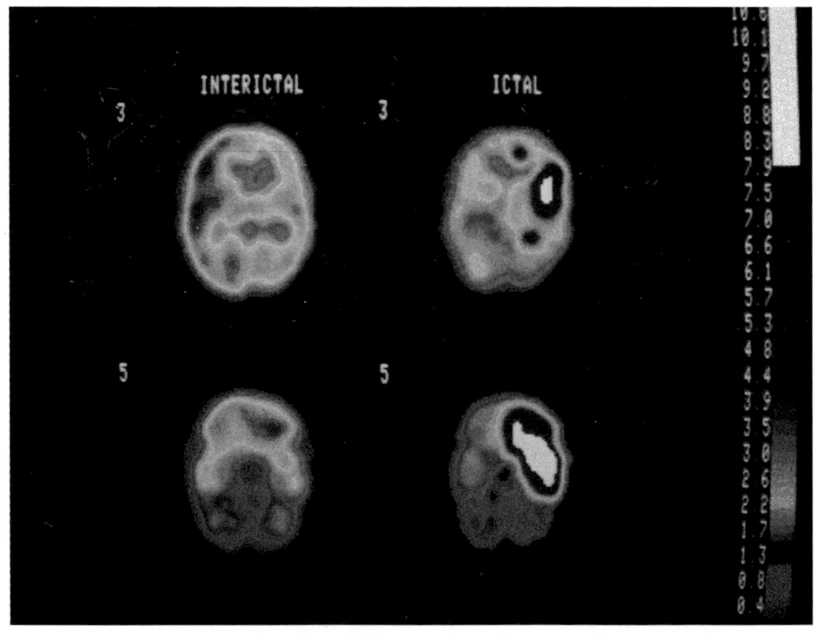

Figure 7.5

PET image of the brain's glucose uptake during the same seizure. (Printed with permission from Engel, et al., Neurology 33:[1983]: 400).

"One type of seizure is related to a 'heat disorder,' involving a very strong pulse, and in this case there is no treatment at all. It is incurable. A second type of seizure is related to a 'cold disorder,' indicated by a more subtle pulse, and that can be cured. An epileptic seizure is seen as stemming from the brain, and two types of major causes are explained. One is due to a swelling inside the brain, and the second is to an external influence, such as tiny organisms that are not visible to the naked eye. These are two causes that lead to seizure. The third type of epilepsy can be determined by reading the pulse: when you press normally, gently, you can feel the pulse, but when you press slightly harder you can't feel the pulse. It's called an 'empty pulse,' for it's empty like a ball: you press it and then suddenly it's gone." He paused and added: "It's very difficult to get a coherent understanding of this without understanding the Tibetan medical system as a whole."

The Dalai Lama offered a bridge: "In my experience, Tibetan physicians can perform diagnoses very precisely without complicated instruments. In my own case, on one occasion I entered a hospital in Calcutta or Delhi, where they used huge machines; but they failed in their diagnosis. Our doctors, without any instruments, touch the wrists, listen to the various pulses, examine the person, then know quite precisely what is wrong. The whole system is really remarkable.

"Here's a general statement about Tibetan medicine: human physiology is spoken of in terms of the three humors—wind, bile, and phlegm. Where do disturbances or imbalances of the wind, bile, and phlegm originate? Wind, bile, and phlegm disturbances come, respectively, from the 'three poisons,' or primary mental afflictions, namely attachment, anger, and ignorance. These three mental afflictions give rise to disturbances of these three humors. Also in some treatises, disorders of the flesh are related to ignorance, disorders of the bones to anger, and disorders of the blood to attachment. Generally speaking, the illnesses of the body are related to imbalances in these three humors, which have their source in the three primary mental afflictions.

"It's also asserted in the Tibetan Buddhist medical system that external agents may give rise to certain physical imbalances. These agents include nonhuman beings such as *devas* (or celestial beings), *nāgas* (or subterranean beings), and other creatures. These are entities dwelling on this planet, and they may injure people, as in Western accounts of possession. Although they may injure people, they're not the major cause of illnesses. Rather, they catalyze an imbalance in these humors.

"There are various forms of diagnosis. For example, there is the diagnosis of the pulse and diagnosis of the urine. Those are both very sophisticated and complicated techniques. There is also inspection of the symptoms of the illness, including, for example, examination of dreams. Once the diagnosis is complete, medication is given. If over a period of time the medication alone is not effective, then additional methods will be introduced, including religious ceremonies. This would not be a general religious ceremony, but one specifically associated with the illness in question. For example, if a physician believes that a *deva* is a factor in the patient's illness, there are specific rituals designed to counteract that influence. Sometimes a patient will experience no benefit after a very long period of taking the medication alone, but as soon as one of these rituals is performed, the medication immediately starts to be effective."

Pete asked: "Is the objective of the medical practice to cure the patient at all costs, or are there times when it's believed that the patient should die and the treatment is stopped?" This question would lead us back into the whole issue of death and the ethics of death.

After a brief moment of discussion with Dr. Choedrak, His Holiness offered a response. "There are precepts that physicians take in the Tibetan Buddhist medical tradition, and one of the precepts states that the physicians are obligated to use whatever means they have available to heal the patient. At no time does the situation arise where physicians believe they have medication that could possibly be of benefit, but they refuse to give it and let the patient die. To do so would entail breaking that precept. Now, of course, it's questionable whether the medication will in all cases be effective; but the physician does have a vow, or a precept, to do whatever is possible to heal."

Indications of Death in the Tibetan Tradition

"What if the patient is unconscious and it's clear they will never regain consciousness?" insisted Pete.

"Based upon pulse diagnosis alone it's possible to distinguish quite clearly certain portents of coming death. In fact, there are multiple ways of examining portents of death. For example, if one is very skilled, one can carefully scrutinize the breath itself, even that of a person who is apparently in perfectly good health, and be able to detect indications of death that may be even several years away. Different types of pulse may indicate relatively distant death, death at a medium distance, or proximate death. In the third case, when death is quite near and the physician knows that there is no more hope, he will not try to raise the spirits of the patient by encouraging him or her to take medication anyway. Rather, the physician will encourage the patient to have whatever food he or she would like. If there were previous dietary restrictions, these are now lifted and no more medication is given. That's because the physician recognizes that nothing would help anyway."

"Are these diagnoses ever wrong?" asked Pete with some amazement. The answer came quickly: "Quite possibly."

"There are stories of accomplished practitioners," intervened Adam Engle, "who will announce that at a certain time they're going to die, and then they die. There is the concept that you can recognize that whatever you come here for during this lifetime is completed and it's time to go. If this is true, at what point would it be appropriate for someone to choose to 'pull the plug,' as we say? There's a woman in California who is not brain-dead but very much alert. However, her body is in such a bad state that she wants to die. She has to be fed and maintained artificially, and the doctors can't stop that. She is suing the state of California to die. It's a terrible situation, but there's nothing that physicians can do in a case like that.

So the question is, when in the Buddhist view is it okay to stop?"

"As a general principle you need to consider what would be of greatest benefit," His Holiness answered. "The wishes of the person who is involved are very important. Then there is the relatives' wish. The cost should also be a consideration. In certain circumstances the patient has no wish to carry on, or has no feeling at all. The most precious thing is a human being. If the brain is not functioning, and if only the body is kept alive at great expense, it would be more useful to spend the money for some other purpose, assuming that there is utterly no hope of recovery. There are cases in which the relatives are willing to spend the money, even when there is no hope of recovery; and that, of course, is their choice. I think from the Buddhist viewpoint, if the person's brain can be used to think, to move, or to increase some positive and useful motivations, such as compassion and so on, this is good. You could also have a case in which the person's brain might be active but accomplishing nothing, in contrast to the person who is willing to utilize that opportunity for enhancing his or her wholesome attitudes, such as developing compassion. On the other hand, you can have cases in which the brain is very active, but the whole brain is used in feeling always depressed, anxious, or worried about one's inability to move, to use the body, and so forth. That would be a different case, in which the brain is used simply to compound one's suffering."

Stages of Death

I was very eager to delve deeper into the stages of dying, so I used Adam's question as a springboard. "In the case of the practitioner who decides to die, sometimes they can stay in a state in which the brain is dead but the body remains alive for some time. We have witnessed people who don't decay for some days. What is your view on that? At what point does death occur?"

"Such a person is said to be in the state of dying, but has not yet entered death."

This was precisely what I wanted to evoke, and Pete helped by adding: "The definition of death seems to be quite different in

Buddhism and in Western medicine. Some time ago I met a Tibetan meditator called Lama Yeshe, who died in California in a Western hospital. The doctors said that he was dead, but his friends said he wasn't and asked that he be left alone. He stayed there for three more days with no decay, presumably in the clear light of death. Finally they said he was dead and moved the body. In that process there is no life going on that can be detected by Western medicine; how does Buddhist medical practice determine that this individual is not yet dead?"

His Holiness turned again to consult with his physician. "There is a fairly simple criterion: you check to see whether the body is decaying or not. If the body is not decaying, then you leave it. However, I have asked Dr. Choedrak whether there's any reference in the medical treatises to abiding in the clear light of death experience, and the doctor's response was that there is not." He reflected for a moment and then continued.

"Let me give a presentation of the dying process and death according to the Buddhist Vajrayāna. This is well established in these texts, but this account has yet to be investigated by scientific means. This discussion focuses on the energy center at the heart, in which there are said to be a very subtle white element and a red element (Skt. *bindu*). In the process of dying, the white element descends from the head, down through the central channel, and then stops at the heart center. From below the heart a very subtle red element, or drop, arises. As the very subtle white element descends to the heart, one has the experience of a pale light. Following this, the red element ascends to the heart, and while this is occurring, there is a subjective experience of a reddish sheen arising. When the two of these completely converge, like two bowls coming together, there is a period of blackout, as if you simply lose consciousness altogether. Following that blackout period is the period of the clear light of death.

"The clear light of death is something that everyone, without exception, experiences, but there is much variation in terms of how long the experience is sustained. For some people it may last only a few seconds, for some a few minutes, for some several days or even weeks. As long as the clear light of death experience is sustained, the connection between the very subtle energy-mind and the gross physical body has not yet been severed. It's in the process of being severed but it has not yet been completely severed. At the very moment that the severance takes place, the body begins to decay, and at that point we say death has occurred. The external sign of this taking place, by means of which one can know with full certainty that death has taken place, is that these red and white elements emerge from the nostrils. This is seen as a red trace and a white trace, which may also be emitted from the genitals. This is true for both men and women."

Gross and Subtle Levels of Mind

I pursued my questioning: "Your Holiness, this kind of account presents a big problem for Western neuroscientists because of its dualistic overtones. The most famous illustration is Descartes, who postulated that the soul and the body interact at one point in the pineal gland. But in its modern guises this problem has haunted Western scientific thought from the beginning. If something is attached to something else, it means that these two things are of different nature. The problem is thus called the *problem of duality*. If mind is different from body, then there is never a way in which these two meet. As scientists we don't like that situation because there is never a way in which one level of phenomena, the mind phenomena, can enter into the biological and physical phenomena. Therefore the use of words such as *severing* is always something that makes us a little uncomfortable. Could you elaborate on how the severing, or in the reverse process the entering of consciousness, can avoid being completely dualistic? Is there a causal connection in this process of severing or entering? If so, what is the nature of this connection?" The answer His Holiness gave to this question was so precise and remarkable that the reader would do well to dwell on it a little longer than usual. It contains a view of mind and body that is neither materialistic nor dualistic in a banal sense. Furthermore, he touches on the very core of the Buddhist meditative experience, beyond intellectual description. At points such as this, the work of setting up these dialogues carefully gives fruits that are sure to make a difference. Listen well. It doesn't get any better!

"To review just a little bit, there's certainly a convergence between Buddhism and science when you speak of gross levels of consciousness. Buddhists would agree that the gross levels of consciousness are contingent upon the body, and when the brain ceases to function, those levels of consciousness do not arise. A very simple example of this would be that if the physical basis for vision is lacking—the visual cortex, the retina, the optic nerve, and so forth—you do not have visual perception. It is very straightforward on that level.

"We've agreed that this consciousness arises in dependence upon the brain. This is a causal relationship, but once again the question can be raised: what is the nature of this causal relationship? Does brain function act as a substantial cause for mental processes, or does it provide cooperative conditions? I have never seen this explicitly discussed in Buddhist treatises, but it would be reasonable to surmise that brain function provides the cooperative conditions for the arising of mental processes. But what is the substantial cause of the primary, distinguishing qualities of consciousness, namely clarity and cognition? The Sūtrayāna system seems to answer that clarity and cognition arise from the latent propensities of the preceding mental continuum. According to Vajrayāna, as stated previously, the substantial cause would be identified as the very subtle mind, or primordial mind. But I understand the heart of your question to be: what is it that provides the connection between the very subtle mind and the gross body or the gross mind? Here we speak of the 'very subtle energy bearing five-colored radiance.' This suggests that this very subtle energy is endowed with the ultimate potential for the five elements: earth, water, fire, air, and space. From this energy there first arise the five inner elements, and from these the five outer elements arise.

"There are some mentalist Buddhist schools, principally the

Yogācāra school, that deny the existence of an external world. One reason proponents of this school posit the foundation consciousness is because they deny the external world. They need this foundation consciousness as the storehouse for imprints that manifest in a dual fashion, both as subject and object. According to this view, you don't need an external world because everything arises from a source that is essentially mental in nature.

"Let's move now to the Prāsaṅgika Madhyamaka school, which we consider to be the ultimate Buddhist philosophical system. This school asserts the existence of an external world, including the objects of the senses. Whence does this world arise? What is the origin of this external world, the physical environment? The origin is traced back to particles of empty space. This origin is not posited as the beginning of all time, for in Buddhism there is no notion of a beginning of all time, but rather the origin within a cosmic cycle. In short, you can trace this whole manifest cosmos back to the space particles. This is not to say that's the ultimate origin of the universe; but that's the beginning for one cosmic evolution. Then you can speak further about what happened before then. The whole of the evolution of the natural world stems from space particles, and that evolution would take place regardless of whether there's consciousness, regardless of the karma of sentient beings.

"This universe is inhabited by sentient beings who experience situations and environments that lead to their detriment or to their happiness. There is an interface between the karma of sentient beings and the natural environment. Karma modifies or influences the nature of the physical environment such that by inhabiting this physical environment one experiences pleasure or pain. In this context we speak of good fortune, misfortune, and so on. What is the source of wholesome and unwholesome karma? This is traced back to mental processes and, more specifically, one's motivation. Wholesome and unwholesome motivations are the most influential factor in determining whether one's actions, or one's karma, are wholesome or unwholesome. As soon as you are concerned with motivation, you're in the sphere of the mind. And the mind is intimately related to the very subtle energy, the energy bearing the fivefold brilliance. This energy bears the potential of the five elements, with the five outer elements evolving from the five inner elements. Thus, karma would presumably have as its vehicle this very subtle energy as it manifests through the outer and inner elements. So there's a two-way interface between the mind and the physical elements."

We were all quite amazed at his exposition, and I barely could retain enough of its richness to clarify: "Does it not follow that my awareness is not hidden but rather embedded, or permeating, the gross mind that depends on the brain—that they must be interpenetrated?"

"That's quite true. It's not the case that you have two distinct continua of consciousness, one very subtle, the other gross. Rather, gross consciousness stems from the very subtle mind; it gets its very effectiveness from the very subtle mind. It's not something separate."

"When we see, for example, a subject listening to sound, there is a lot of very complex brain activity. Do you think, Your Holiness, that from an external picture of this we might be able to see not only the gross but also the subtler levels of consciousness?"

He thought for a moment and then said: "It would be very difficult to determine by scientific means. A contemplative who is very highly advanced in his practice, and who has gained a direct experience of the very subtle mind, needs no outside scientific proof to ascertain the existence of the very subtle mind. But without that experience, the existence of the very subtle mind can't be proven. This can be compared to an analogous situation discussed in Mahāyāna treatises that mention certain signs of *irreversibility*. This is a specific stage on the spiritual path. Once one has reached this stage, one never regresses on the path, one only progresses further. The question is then raised: how can we determine whether a person possesses these signs? There are two conflicting positions on whether this irreversibility can be conclusively determined at all. Even those who maintain it can be proven admit that the proof is not one of inference, but rather like a proof by analogy. The example given is that you can verify the identity of a particular house by saying that this is a house on which a black crow was standing. You use the crow as a sign indicating the identity of that particular house. This is indirect evidence, which has nothing really to do with the nature of the house. Nevertheless, it helps you to recognize that particular house. The signs indicating the stage of irreversibility are to be understood in a similar fashion.

"To determine whether the innermost subtle clear light exists or not, we can proceed in the following way. When we observe the mind, we know from experience that there seem to be three fundamental ways of responding to situations or events. One is repulsion, one is attachment, and one is a state of indifference. Among these three states, it is said that repulsion requires the strongest energy, attachment requires less, and indifference even less."

He turned to me: "By means of EEG, have you noticed a difference for a person who is experiencing strong anger, as opposed to strong attachment or desire? Are there any quantitative differences?"

"Your Holiness, the EEG is a very gross measurement. If one does more elaborate computations and data analysis, some patterns can in fact be detected. In particular, there are some reliable relations between strong emotions and the putative sources of surface recordings. As of now it's still a very open area, but it's not a problem in principle. It is technically difficult, but the skillful combination of EEG, magnetic field recording (MEG), and the new brain imaging techniques of PET and functional magnetic resonance (FMRI) hold great promise."

He resumed: "This is related to my point concerning the differences in energy required for repulsion, attachment, and indifference. In Nāgārjuna's writings it is said that there are eighty types of conceptualization (Skt. *saṃskāra*), which indicate various levels of energy activity, and these are associated with various states of emotions and thought. They are divided into three groups based on the level of energy activity: highest, medium, and lowest, respectively. The first group has thirty-three, the second has forty, and the third has seven types of conceptualization.

"On what grounds do we postulate the existence of the clear

light? At least we have to present a plausible case. The eighty types of conceptualizations are various emotional and cognitive states that are elements of the mind. These eighty are said to cease functioning when respiration stops. This could also happen when brain function stops. According to this system, there are three fundamental states from which the three divisions of conceptualization arise, and these three primary states are called *appearance*, *increase of appearance*, and *blacking out*.

"According to one theory, these three states themselves must arise from the basis of the clear light state, the innermost subtle state of consciousness. If you carefully investigate the treatises that present this theory, and if you investigate points that do lend themselves either to experiential confirmation or cogent inferential confirmation, you may come to have confidence in the veracity of those states that do not lend themselves to such confirmation. You have two alternatives: either to dismiss them, or accept them since there doesn't seem to be any contradictory evidence.

"To see how this approach works, we have to understand a threefold Buddhist classification of phenomena: (1) *evident phenomena*, which are known through direct perception; (2) remote or *obscure phenomena*, which are known through inference; and (3) *extremely remote* or *obscure phenomena*, which are known only through the testimony of a third person. Such knowledge based on another's testimony occurs within the context of a system in which you have developed a high degree of confidence because of your own investigations. If you have gained that trust yourself through your own striving, then within that context it is fitting to speak of an inference based on testimony. This is very different from simply accepting another's testimony without any such context."

"But the basic framework is that we avoid two extremes," I ventured. "The dualistic extreme would be that clear light and gross consciousness are completely separate. The other extreme would be that clear light and gross consciousness are completely mixed. The intermediate position here seems to be that clear light is the source of continuous levels of manifestation, the deepest of which are very hidden."

"Not quite like that," came the reply. "It's not simply a causal relationship in which one phenomenon gives rise to another distinct phenomenon. To start with, the very subtle mind and gross mind are of the same nature, not of distinct natures. This whole topic seems to be clearest in the Dzogchen literature. Here the relationship between the innermost clear light and gross consciousness is not simply dualistic, but is treated with far more subtlety. As I mentioned yesterday, the clear light of death, which manifests at death, is also called *natural pristine awareness* (Tib. *rig pa*). However, even while gross consciousness is manifest, it is also possible for pristine awareness to manifest, but now it's given another name. As mentioned yesterday, it's called the *effulgent pristine awareness*, or *pristine awareness involving the appearance of the basis*. This manifests simultaneously with gross consciousness, so it's not the case that pristine awareness is completely dormant as long as gross consciousness is manifest. The fact that one can experience the effulgent pristine awareness while there is still gross consciousness suggests that the former is more pervasive, rather than being simply a cause of gross consciousness."

He paused and pondered with a gentle astonishment. His tremendous appreciation for the Dzogchen tradition of meditative practice was clear. "I have questioned certain Dzogchen practitioners who are cultivating this experience of pristine awareness, and asked them about their experience. In particular, I recently met a 25-year-old contemplative who has had very clear experience of pristine awareness. Not too long ago he returned to Tibet, to the region of Dzamthang. I asked him about his experience of clear light, specifically during waking consciousness. The contemplative said that right in the midst of his waking consciousness he was able to ascertain that very facet of pristine awareness itself. In general, we speak of the nature of awareness as consisting simply of clarity and cognition. When one has manifest experience of pristine awareness, one can ascertain these very facets of clarity and cognition, distinct from the clarity or cognition of a specific object.

"This is the report of people who have excellent experience in Dzogchen. Through that experience you also gain more of the actual experience of clear light. Even though you do not have experience of the clear light of death, nevertheless by following this practice you're approaching a deeper and deeper experience of that clear light."

Gross and Subtle Sexual Intercourse

"Since it can only be ascertained by indirect evidence, one could never use gross scientific measurements to ascertain that pristine awareness exists in the experience of a living person," I insisted, unable to resist the neuroscientist inside.

"It's quite possible in principle that you would find something remarkable if you were to apply certain types of research to such people when they are abiding in meditative equipoise. We must distinguish this from the *basic* pristine awareness. I do not feel that you can find scientific evidence for the existence of the basic pristine awareness, but that's not what we're speaking about here. Here we are discussing the *effulgent* pristine awareness, and it's quite feasible that its existence might be indicated by means of scientific research.

"For instance, there's a great difference between the movement of the regenerative fluids for two individuals engaged in ordinary sexual intercourse as opposed to a highly realized male yogi and female yogini who are engaged in sexual intercourse. Although there is a general difference, there should be similarities from the time when the regenerative fluids begin to flow down until they reach a certain point. In both ordinary sexual intercourse and in the sexual union practiced by advanced tantric practitioners, the regenerative fluids move to the point of the genitals. Because of this it would be possible to conduct research to learn about the processes occurring in the ordinary sexual act.

"In principle, the general difference between the two types of sexual act is the control of the flow of regenerative fluids. Tantric practitioners must have control over the flow of the fluids, and those who are highly experienced can even reverse the direction of the flow, even when it has reached the tip of the genitals. Less experienced practitioners have to reverse the direction of the flow from a higher point. If the fluids descend too far down, they are more difficult to control.

"One training method that can be used as a standard of measurement of the level of one's control entails inserting a straw into the genitals. In this practice the yogi first draws water, and later milk, up through the straw. That cultivates the ability to reverse the flow during intercourse. Those who are highly experienced can not only reverse the flow from quite a low point, but they can draw the fluid back up to the crown of the head, from which it originally descended.

"What actually is the white element that is drawn up all the way to the crown of the head? According to Western medicine, the sperm comes from the testicles, and the semen comes from the prostate. According to Tibetan medicine, the semen comes from the seminal vesicle (Tib. *bsam se'u*). What's being raised up here? Is it the semen, the sperm, or something else? Dr. Tenzin Choedrak has corroborated that it's neither of these gross substances, both of which are accounted for in the Tibetan medical system. Rather, it is a very subtle substance that is actually drawn up to the crown, not the gross fluid of either the semen or the sperm.

"One may then ask how it gets up there. Through what channel, or by what means does it move? There are three channels—the central, right, and left channels. There are six centers on the central channel, all of which have knots in them that must be unraveled. Until one has reached the very highest state of practice by loosening all the knots in the various centers, there's no way that this white element can pass through. There must be a clear thoroughfare through all these centers before the white element can go through. For an advanced yogi who hasn't reached the highest state, the white element is drawn up to the crown through the left and right channels. When the yogi gets to the highest state of practice, the white element passes through the central channel to the crown of the head. Once you've unraveled the knots, they stay that way.

"For women, these six centers are the same as for men. It's said that the red element is more dominant for women, but they also have the white element. The fact that women have the white element is one more indication that this does not refer either to semen or to sperm. I have spoken with some Hindu yogis who are very experienced in *prāṇa* yoga and in dealing with channels and energies. Some of these people have reported witnessing that the white element is also present in women, although the red element is stronger. So for women practicing the tantric meditation described earlier, the white element descends in exactly the same way and is drawn back up again. In tantric literature, four types of women, or consorts (Skt. *mudrā*), are discussed. These four types are lotus-like, deer-like, conch-shell-like, and elephant-like." Very aware that the topic was raising smiles all over the room, he joked: "If the classification had originated in Tibet instead of India, they would have called it yak-like." We all laughed with some relief. "These distinctions all have primarily to do with the shape of the genitals, but they also refer to differences in terms of bodily constitution. There are no such categories for men.

"You see, it's subtle. And at the time of death, this process is very difficult to investigate."

Transference of Consciousness

"If one has a very powerful practice of consciousness then the evidence is very clear. One thing that we can learn from the practice of *powa*, or transference of consciousness, is the effect of consciousness on the body. If you have become very skilled at this, when you practice *powa* the body falls over, even though you're in perfectly good health. Sometimes there's a swelling on top of the head, and some emission of fluid due to this mental practice. Because of this, the tradition recommends that the practice of *powa* should be followed by practices for prolonging life. There are some cases of Tibetan Buddhist contemplatives who did this practice while they were being brought to Chinese prisons in Tibet."

Pete asked: "It seems to me that *powa* would be an excellent practice on which to conduct the kind of research that you're talking about, because there are physical manifestations of the subtle mind. The reservations you have expressed about doing experiments on individuals in the process of dying would not apply. *Powa* provides an opportunity for such research without interfering with the dying process of an advanced practitioner. In this context, could you briefly describe *powa* to those of us who don't know what it is?"

"You use meditative visualization to sever the connection between the subtle mind and the gross body, without any damage to this body. If one practices *powa* before one has any portents of death and without any proper reason, there's a danger that you might unintentionally commit suicide. However, you might detect portents of approaching death even while you are in good health. These signs may be seen three or four months prior to death. This justifies practicing *powa* and ending your life three or four months early, inasmuch as letting your body deteriorate in sickness would make it very difficult to meditate and exit from this life properly.

"When a meditator applies the technique of transference of consciousness, or *powa*, to sever the link between the gross body and the very subtle mind, then that is actually an experience of death. Although you may not go through the stages in a prolonged way, you do go through them gradually, in their proper sequence, culminating in the actual experience of death.

"There is another type of practice called *drongjuk* (Tib. *grong 'jug*), in which you send your continuum of consciousness into another fully formed body. This other body is not a living body, you're not pushing anybody else's consciousness out or killing anybody. Rather, you insert your consciousness into a fresh corpse. This is the equivalent of a body or brain transplant: the second body becomes the first person. It is said that people who do this can carry with them all the skills that they have learned. They do not actually experience death, since they haven't gone through the eight stages of dissolution.

"But bear in mind that in Buddhism suicide is unwholesome. *Powa*, as I mentioned earlier, should be applied only when you see the portents of death. Untimely application is equivalent to suicide. In proper practice, you must have determined that you're going to die soon; then you accelerate the process while you are strong, and that's acceptable."

Experimental Occasions for Subtle Mind

Given the depth of the issue, we all seemed to have a hard time drawing away from this possible bridge between the teaching on subtle and gross mind and neuroscientific tools and methods. Seeing that time was quickly running out, I tried to summarize: "Of all the conditions that allow us to ascertain the subtler levels of consciousness, the clear light of sleep and the foundational pure awareness described by Dzogchen practitioners would seem to be the most directly available to Western medical science."

His Holiness agreed: "You can also conduct research on someone doing *powa*. Virtually no one does the practice of transference of consciousness into a corpse. This tradition is being lost."

It was Pete's turn to insist: "But you would expect there to be physical properties of the body in those states that are similar to the clear light of death?"

His Holiness added: "A fourth possibility for research would be to take measurements on meditators practicing *vase breathing*. Meditators who can hold vase breath at the lowest level of realization are said to be able to hold the breath for just over two minutes. More advanced meditators can last four or five minutes, or even eight to nine minutes without breathing. There is a heartbeat, perhaps. According to medical science, is it possible to hold the breath this long? How many minutes do experienced Japanese divers and others remain under water without breathing?" The answer from the two physicians present was the same: about five minutes. He insisted: "Is it possible, according to medical science, to spend nine or ten minutes without breathing?"

"Yes, by lowering your metabolic rate," said Pete. "You can do it if you cool the body, but a practitioner might have other ways of decreasing the metabolic rate to protect the brain from a lack of oxygen. I think the PET scan could answer some of these questions, but it would require injecting the tracer into the vein. Would these practitioners allow that to be done?"

"That is an individual prerogative. They might not like it right while they're meditating. It's possible it could cause some disturbance, but that most probably depends on the practitioners' level of experience. If they have genuine, deep, stable, and subtle experience, I think there would be less disturbance."

Pete expressed what we all felt by saying that such experiments would be remarkable. With his open-mindedness towards science, His Holiness offered: "I'll seek people for that. Our task is to find the subjects. You scientists can produce some terrible machines, and we'll try to find a dead person first." We all had a good laugh. "In the sixties I met a scientist, who is my dear friend, and spoke about these phenomena. He told me he would like to conduct certain types of research, but I told him the meditators he would need to work with hadn't been born yet! Over the past twenty years, there have appeared some people with experience, but it's still difficult to find them, for they are scattered. In Ladakh, there are some truly religious Buddhist practitioners, but they are completely independent. Nobody can tell them what to do. Some are quite stubborn. Anyway, I believe this is something very important, if you take a broad view.

"We have already discussed one of the main premises on which science and Buddhism can have a dialogue, namely the Mahāyāna position that encourages individuals to be open-minded. This entails adopting a critical stance towards the Mahāyāna teachings themselves. There must be an active engagement and questioning for yourself, not just disinterested skepticism that may imply that you don't take the subject seriously. Buddha said that his words should not be accepted simply out of reverence for him, but rather they should be examined as a goldsmith examines the gold that he's going to buy. So that is the basis of our position: investigation, more investigation, more discussion." He chuckled softly. It was the perfect closing remark for a remarkable session.

8

Near-Death Experiences

Death as Rite of Passage

JOAN HALIFAX IS, IN TRADITIONAL categories, a cultural anthropologist. She began her career in cross-cultural anthropology, then moved to medical and then clinical anthropology. Now she finds herself primarily known as a cultural ecologist. She lives in the midst of change and thrives on it. Her sharp blue eyes revealed her usual intensity as she sat next to His Holiness.

"What we're looking at today are the so-called near-death experiences. These are the stories that people recount of their experiences after going through clinical death and being resuscitated, or spontaneously reviving.

"The mode of story is quite interesting. In anthropology we recognize that the event-filled story is a way that human beings across all cultures weave cosmological insights, or insights about the nature of the self, into the fabric of society. If we look at these narratives cross-culturally, we discover that the story of the event of death and the experience immediately following is extremely common. It appears in Western culture, for example in medieval accounts of near-death experiences; in Eastern accounts such as the *Bardo Thödol*,[21] where the death experience is very well explicated; and in tribal societies everywhere in our contemporary world. Death is not only expressed in stories, but is also clearly acted out in culture. There are many rites or ritual events that conduce an experience of death and rebirth.

"These ritual events, which are commonly called *rites of passage*, happen not only periodically in an individual's life, but also in terms of geographical transit, such as a journey into exile. A rite of passage, in other words, is a ritual event that is about dying relative to the old way of being and about being reborn into new understanding, a new way of life. It can be predicated on age: for example, adolescents go through a rite of passage. Women giving birth go through a rite of passage. Marriage is a rite of passage, and the relatives of deceased people go through a rite of passage. There are rites of passage associated with the experience of maturation. In many cultures these rites are not superficial events in human experience. In tribal societies, they are frequently effected rather dramatically. For example, young boys going through a rite of passage marking adolescence may enter a period of extensive seclusion and face body mutilation. They may be fed stories or myths that give a cosmological grounding to the experience, preparing the adolescent for adulthood. A series of events might take the individual into an altered state of consciousness, where normal understanding is disordered or even destroyed. An initiate may even go into a coma or something like a near-death experience, and subsequently revive with the intent of experiencing some kind of illumination.

"Rites of passage are events that not only prepare an individual for life; they also prepare an individual for death. These rites do not exist per se in Western culture. The absence of rites in Western culture is very consequential, resulting in alienation. Death is repressed in Western culture, as Charles already mentioned. However, by the time the average teenager in the United States is eighteen years old and graduates from high school, he or she has witnessed over twenty thousand television homicides. Death not only emerges in ways that are unhealthy for society—the repression giving rise to a kind of obsession with death—but there is also a conditioning that objectifies and alienates the experience of death from the kind of intimacy and compassion that is part of Tibetan practice."

Exploring the Edge of Death

"In the United States during the 1960s, the war in Vietnam and the Civil Rights movement gave rise to a deep impulse to liberate peoples who were socially and culturally oppressed. Here was a social and mental revolution in which people did a great deal of experimentation with spiritual practice. At the end of the sixties, death and dying became an area of spiritual, psychological, and philosophical exploration.

"I was part of that pioneer wave in the early seventies when I participated in a project sponsored by the National Institute of Mental Health. In that project a group of psychiatrists and social scientists worked with mind-altering substances in relation to people who were dying of cancer and individuals who were suffering from acute pain, fear, or depression. We interviewed individuals who were recommended into our program and then entered into a very profound psychological interaction where, by consent, they were given a psychoactive drug in the course of very dynamic psychotherapy.

"Psychoactive drugs are substances that have been used by many tribal cultures all over the world for their capacity to induce a profoundly altered state of consciousness. This category of substances began to be investigated in the late 1950s in Europe and then in the United States. This particular project used LSD, one of the first man-made psychotropic substances.

"A number of things became very clear to me as an anthropologist. First, this work seemed to be a useful experience for people who were not prepared for death—a contemporary rite of passage. The second was that death itself involves an alteration of consciousness—many psychological effects contribute to the transformation of mental conditions—and that it probably wasn't necessary to use a mind-altering substance.

"Since that time I've continued a practice of clinical anthropology that involves sitting with people who are dying. I teach dying people meditation and encourage them not to shy away from the altered states that come in the course of the experience of dying, but to establish a strong mental foundation so that they can investigate these states with some degree of equanimity. For the past six years I've been working with homosexual men who are dying of AIDS.

This population is particularly interesting: these men tend to be well educated and quite spiritual. They see the experience that they're going through as a mission to their brothers, many of whom are suffering from a similar condition. They are motivated to do a good job in the dying process."

Archaeology of Death Rituals

"If we turn to human origins, we discover a long record of human attitudes toward death. Bodies are interred with artifacts, or placed in certain positions. For example, severed heads were found at a burial site half a million years old in a small cave in China. Paleoanthropologists conclude that there were practices either of ritual cannibalism, or of reverencing the head as the seat of consciousness. Among Neanderthal people of the Middle and Near East, sixty-thousand-year-old archaeological sites yield a rich fossil record. There is a cave in Le Moustier in the south of France, where the remains of a teenage boy were found in a pit, with his head resting on his arm in the position of sleep. With him a stone ax and a funeral meal were found. Paleoanthropologists have surmised that the burial indicates belief in an afterlife, or a journey in the afterworld. We see this also in tribal cultures that have survived since the Paleolithic period sixty thousand years ago.

"One of the most fascinating burials is in Shanidaar in Iraq. The corpses were laid in a bed of flowers, many of which have medicinal value. It is believed that these flowers were somehow part of the resources that are taken into the afterlife. During the Cro-Magnon period, from about 35,000 to 10,000 years ago, people were buried bound or bent into the fetal position, making the very close connection between the experience of death and the experience of birth. In the Upper Paleolithic, 15,000 years ago, mostly in France in the Dordogne, we find cave paintings indicating ceremonial behaviors and attitudes around death that we might call sacred. The most interesting painting is in the cave of Lascaux, where there is a man in a supine or a reclining position. The man appears as if dead, but has the mask of a bird on his head and his penis is erect. Beside him is a bison that has been speared in such a way that his entrails are falling out. The dying bison turns his head as if to look at his own entrails and also seems to be gazing back at the man. We believe that there was a relationship in the minds of Paleolithic people between birth, death, sex, and trance. It seems as if the practices that contemporary tribal peoples have carried through the centuries since the birth of man have something to teach us."

Western Discovery of the Afterlife

"In Western culture, there is a particularly vivid time of the investigation of death during the medieval period, from about 500 C.E. until around the fifteenth century. There are numerous accounts of near-death experiences by a wide range of people: rich and poor, popes, kings, and children. These testimonies were oriented towards validating the afterlife. Visionary testimonies like those given in the Middle Ages had to do with reinforcing proper behavior in this life to avoid hell in the afterlife. Interestingly, in current near-death research into accounts of death in contemporary culture, hell is not a prevalent theme.

"Death again becomes a subject of great interest in the late 1800s. A man called Albert Heim, who was a geologist and a Swiss mountain climber, fell off a mountain and had a powerful mystical experience. He developed a deep interest in individuals who survived accidents, and began to collect accounts of near-death experiences. These stories came from individuals who were not particularly spiritually inclined, but had had spiritual experiences in the context of their accidents.

"Also in the late 1800s, spiritism arose. Spiritism is the ability to see and communicate with spirits of the dead. The spiritist movement became widespread in the early 1900s, and with its popularity a new area of research emerged, exploring paranormal or psychic events. A number of books were written on the human personality and its survival of bodily death. Key researchers at this time were F. W. H. Myers, James Hyslop, and William Barrett. In 1918 one researcher proposed a census of death bed visions to support the claims of spiritists.

"By the 1950s this sort of psychic and spiritual bent became a little less fashionable and the area that is now called parapsychology, or extranormal states of consciousness, opened up. In the late 1950s, the researcher Carlos Osis began to collect accounts of near-death experiences of physicians and nurses. Because comparative religion was of interest then, Osis and his colleague Erlendur Haraldsson began to study near-death experiences in India as well. Another parapsychologist, Ian Stevenson, of the University of Virginia, was especially interested in verifying reincarnation experiences.

"In the late sixties parapsychology wasn't so fashionable. An area of Western psychology called transpersonal psychology was developing. It was not specifically psychic but more related to the work of William James (1842–1910), a philosopher and a student of mystical experiences. The emphasis shifted to the value of these experiences in people's lives.

"In the sixties and seventies, the psychologists Russell Noyes and Roz Kletti at the University of Iowa began to study near-death experiences as a pathological syndrome. In the early seventies, the psychiatrist Stanislav Grof and I did research with people dying of cancer. This was followed by the work of Elisabeth Kübler-Ross, who saw dying as a rite of passage, and the work of Raymond Moody on near-death experiences. By the mid-1970s, Your Holiness, there was a great interest in death."

Testimonies and Their Patterns

"I now want to give you brief accounts of people who have been through near-death experiences. I also want to stay within the realm of discrete, detailed elements. The *Bardo Thödol* is all grounded in detail. What I'm going to be giving here is not so extraordinary compared to what you have unearthed in your investigations, but I give it for comparison to see if we together can construct a picture of the actual experience of dying.

"This is an account that a certain Dr. Richey gave of his near-death experience. In the first minutes of the crisis he found himself outside his body, staring at the dead form which he could only recognize by a fraternity ring on one of his fingers. He was so disturbed that he fled the hospital in this disincarnate state. He headed toward Richmond, Virginia, to keep an appointment, but he was so distressed that he soon returned to search for his body. When he finally found himself again, he was not able to reenter his flesh. At that moment, the room filled with light and he felt as if he were in the presence of Christ, who directed him to review all the deeds of his life. Christ then took him on a tour of various regions of suffering and bliss, realms vastly different from our own world, though they seem to occupy the same space. After a brief glimpse of glowing streets, buildings, and shining throngs, Richey fell asleep and woke up in the hospital room. Richey was convinced that he had been returned to life 'to become a physician so that I could learn about man and then serve God.'"

As if relieved that the background was complete and the actual empirical observations had begun, His Holiness asked how long the person had remained clinically dead. "One of the things that is constant through all the experiences is the alteration of the perception of time," Joan replied. "Events happen in a very short time. It seems as if an entire life can be reviewed by a mind in a moment. In any case, Richey had had a heart attack and was under reanimation when all this happened."

The Dalai Lama was silent for a moment and then added, "It would be difficult to determine whether this person actually was out of the body, or whether the individual simply had a fantasy of being out of the body."

Pete Engel rejoined, "But your criteria from the other day would pertain: if they were really out of the body doing something else, they could report things when they woke up that they wouldn't have known otherwise."

"That's coming later," Joan said. "First let me point to some patterns that emerge from the testimonies like that of Dr. Richey. In figure 8.1 you find patterns of near-death experiences from three main contemporary researchers: Raymond Moody, originally a philosopher and now a psychiatrist; Kenneth Ring, a psychologist; and Michael Sabom, a cardiologist.[22] We see here prototypical sequences of events through which a person going through a near-death or clinical death experience can pass. The italicized phrases are common to all three researchers, but there are interesting variations in the actual sequence of events that unfolds.

MOODY	RING	SABOM
hearing the news		sense of being dead
feelings of *peace*	*peace*	emotions: *peace*, rest
the wise		
dark travel	*body separation*	*body separation*
out of the body	entering darkness	
meeting others	presence	observing physical events
being of *light*	life *review*	
the *review*		
the border	decisional crisis	dark region
coming back	the *light*	life *review*
talking to others		*light*: blissful beings
effect of individual	return	
new view of death		*encountering others*
corroboration	*talking to others*	

Figure 8.1

Near-death experience structural narratives according to three independent studies.

"In Raymond Moody's account of a prototypical near-death experience, a man is dying and as he reaches the point of greatest physical distress, he hears himself pronounced dead. He then begins to hear an uncomfortable noise, a loud ringing or buzzing, and at the same time feels himself moving rapidly through a long dark tunnel. After this, he suddenly finds himself outside of his own physical body, but still in the immediate physical environment. He sees his own body from a distance as though he is a spectator looking down. He watches the resuscitation attempts from this unusual vantage point and is in a state of emotional upheaval, but we don't yet know why he's upset. After a while, he collects himself and becomes more accustomed to his odd condition. He notices that he still has a body, but one of a very different nature and with very different powers from the physical body he has left behind. Soon other things begin to happen. Others come to meet him and to help him. He glimpses the spirits of relatives and friends who have already died, and then a loving warm spirit of a kind he has never encountered before, a being of light. (Such a being of light is not usually incarnate or anthropomorphized, but it can be.)

"This being of light asks him a fundamental question. It's the question of all questions, but we don't necessarily know what it is, as it's usually presented nonverbally. This question forces the subject to evaluate the events of his life. This is what we call the life review. It is very interesting that, unlike in the medieval accounts, the sense of judgment or guilt does not exist for these subjects.

"At some point the subject finds himself approaching a sort of barrier or border, apparently representing the limit between earthly life and the next life; yet he finds that he must go back to earth, that the time for death has not come. At this point there is generally resistance. People don't want to go back, interestingly enough, for by now the subject is taken up in his or her experiences in the afterlife and doesn't want to return. He is overwhelmed by intense feelings of joy, love, and peace. However, despite his attitude, he has to reunite with his physical body, and he lives. Later he wants to tell other people; there is a messianic desire to communicate what was understood in the afterlife. Most people in Western culture respond negatively to death and so they internalize related experiences and don't bring them out in communication to others. But these subjects feel that their life has been transformed by this mystical experience happening in the context of near death."

"Does the age of the person who's having the near-death experience make any difference?" His Holiness asked.

"All these researchers worked with a sample of individuals that included children as well as old people. Ring's studies include about 150 accounts, and at least that many appear in Moody's. Sabom has 34 subjects in his study. His criteria were more rigorous, though his results were basically the same."

His Holiness continued, his usual skeptical self. "Have many of those who have had near-death experiences ever encountered the *Bardo Thödol*, so that they might have been influenced by what they've read?"

"No, absolutely not. There are some symbolizations of familiar religious characters, for example Christ and the saints, but actually they don't show up that often. As a matter of fact, things happen to people that they don't believe could ever happen."

His Holiness persevered, "But you can also have people who at a very early age have had some religious input. On a conscious level they might be totally atheist, but at a very deep level there may still be religious impressions."

"Yes, of course, but always in Western terms," Joan insisted. "So this is Ring's description," she continued. "The experience begins with a feeling of easeful peace and well-being, which soon culminates in a sense of overwhelming joy and happiness. The ecstatic tone, although fluctuating in intensity from case to case, tends to persist as a constant emotional ground as other features of the experience begin to unfold. At this point the person is aware that he feels no pain, and does not have any other bodily sensations. Everything is quiet in the beginning. These cues may suggest to him that he is either in the process of dying or has already died.

"He may then be aware of a transitory buzzing or windlike sound, but in any event, he then finds himself out of his physical body, looking down at it as though from some external vantage point. At this time he finds that he can see and hear perfectly. Indeed, his vision and hearing tend to be more acute than usual. He is aware of the actions and conversations taking place in the physical environment. He finds himself in the role of a passive, detached spectator watching the drama happening. All this seems very real to him, even quite natural.

It does not seem at all like a dream or a hallucination. On the contrary, his mental state is one of clarity and alertness.

"At some point, he may find himself in a state of dual awareness. While he continues to perceive the physical scene around him, he may also be aware of 'another' reality and feel himself being drawn into it. He drifts, or is ushered, into a dark void or tunnel and feels as though he's floating through it. Although he may feel lonely for a time, the experience here is predominantly peaceful and serene. All is extremely quiet, and the individual is aware only of his mind and the feeling of floating. All at once he becomes sensitive to, but does not see, a presence. The presence, who may be heard to speak, or may instead merely induce thoughts into the individual's mind, stimulates him to review his life. The presence asks him to decide whether he wants to live or to die. The review of his life may be facilitated by a rapid and vivid visual playback. At this point he has no awareness of time or space; the concepts themselves of time and space are completely meaningless. Neither is he any longer identified with his body or, for that matter, any body. Only the mind is present and is weighing logically and rationally the alternatives that confront him at this threshold separating life from death. The question is whether to go further into this experience of death or to return to life. Usually the individual decides to return, but not of his own preference. He is constrained by the perceived needs of his loved ones.

"Sometimes, the crisis of the decision to return or stay occurs later or is altogether absent. The individual may then undergo further experiences. He may, for example, continue to float through the dark void toward a magnetic and brilliant golden light from which feelings of love, warmth, and total acceptance emanate. Or he may enter a world of light and great beauty to be temporarily reunited with relatives who are already dead. In effect, these relatives will probably tell him it's not time for him to die. He has to go back, and so he does. Once the decision is made, the return is usually extremely rapid. Typically, however, he has no recollection of how he effected his reentry, for at this point he tends to lose all awareness as he comes back to life. Very occasionally the individual may remember returning to the body with a jolt or even an agonizing, wrenching sensation. Or he may even suspect that he enters through the top of the head.

"Afterwards, when he is able to recount his experience, he finds that there are simply no words adequate to convey the feelings and quality of awareness he remembers. He may also become reticent about discussing it with others because he feels nobody will understand or they'll think that he's crazy.

"In the last report by the cardiologist Michael Sabom and his associate Sarah Kreutziger, much more rigorous criteria were used to determine whether individuals had gone through a near-death experience. But they came up with a configuration similar to those of the other two researchers. What is interesting about their report is not necessarily new information, but the fact that in several cases they had verified that the person was pronounced clinically dead. It surprised them and increased their enthusiasm for the research because they actually hadn't believed that it could happen. In a large survey the Gallup poll found that fifteen percent of the population in the United States has had a close brush with death and of those, thirty-four percent had feelings of peace and painlessness, an out-of-body experience, or a sense of being in another world and experiencing the life review." Joan concluded by asking His Holiness whether any of this material seemed familiar to him.

Detailed Nature of Near-Death Experiences

At this point His Holiness offered the first challenge to the current interpretation of these observations as related to "true" death experiences. "I am wondering whether these experiences are more of a dream type, because one constant theme that keeps coming up is this experience of joyfully reuniting with the relatives. It is very rare that relatives who have already passed away will still be in that type of existence. They would have already taken rebirth in another realm of existence. There's just the barest possibility that those deceased relatives would still be in the state where they could contact people who were having their near-death experience. But it's almost an impossibility. I think that it's more likely that due to the manifestation of a person's own imprints, or latent propensities, these very images of loved ones come to mind and one has the sense of receiving advice or encouragement from them. But it's purely a subjective phenomenon."

Bob Livingstone remarked that that would agree with the fact that in the West individuals believed to be in Christian heaven and not reincarnated would be available to relatives undergoing near-death experiences.

His Holiness asked for clarification here. "According to the view that you go straight to heaven or hell following death, what's the significance of the final judgment? If you've already made it to heaven, then who cares about the final judgment? And then there's not really any point, is there, in preserving the body in a coffin? The idea, as I've understood it, is that when the final judgment occurs, there is a resurrection of the body."

Charles Taylor clarified, "This is the difference between the Jewish view of resurrection and the Christian view. In the Jewish view, when you die you remain in the grave and when the Messiah comes you are resurrected. In the Jewish tradition the Messiah has not yet come. In the Christian tradition, the Messiah has come already, and death has been conquered so you can be resurrected into the afterlife immediately upon death. The whole thing has to be understood in the context of this paradoxical relationship between two kinds of time. If you view eternity as another dimension, you can't unparadoxically date the last judgment. So there is a deviation from what we would expect, as Western Jews or Christians, in contemporary near-death experiences: there's no bifurcation of the path, no choice about going to heaven or hell."

"I am still trying to figure out whether these experiences are actually dreamlike experiences," he persisted.

I proposed to use the test he himself indicated. "What about the evidence that a patient could recount what was being done to him by the doctors? Would you accept that as evidence?"

"Even in that case, one still has to determine whether that might be an out-of-body experience during the dream state, in which one takes on a dream body, or whether it is an actual out-of-body experience. In other words, is it a dreamlike experience while you're still living, due to the influence of an illness, or is it a genuine *bardo* experience? That's still to be determined. Maybe there's a third option in addition."

Joan said, "Let's look at some of these detailed characteristics and see if you can tell. You're probably one of the very few people who can judge the evidence. One researcher said that fifty-eight percent of the people they worked with had a brand new body: it was the same size and age but of a lighter weight. Another researcher found that the body seems to take the form of some kind of cloud or a sphere that mimics the form of the physical body but is transparent and mist-like. Another researcher said that the body is exactly like the one that we have now, but it's free of all defects."

His Holiness responded, "In Tibetan Buddhist literature the body of a *bardo* being is explained as being free of all defects, even if the person in the former life had physical deformities. There are two main positions concerning the exact nature of this physical form of a *bardo* being. One says it is similar to the previous life, and the other says it is more like the body of the life to come. A third position says the body during the first half of the *bardo* resembles one's previous body, and during the second half it resembles the body in the future life."

Joan added, "The final theory about the body of the soul is that in fact there isn't any: you don't experience a body at all."

"Do you have a feeling of formlessness then?" His Holiness asked.

Joan assented, and added that in some accounts there is a cord which individuals experience as attaching them to this body. At the exact moment of death, whenever that might be, it is severed. "The cord appears only in a few accounts. It's a detail that could very well relate to birth imagery."

"This is a little bit like a common visualization practice in which you imagine Maitreya," said His Holiness. "Then from the heart of Maitreya something like a cord of cloud is emitted, coming down to another being with whom you have a very special relationship. This could be, for example, LamaTsongkhapa. This may be somewhat analogous to what you have described."

Feelings and Sensations

Joan continued, "Next let us examine thoughts, feelings, and sensations in the after-death state. First of all, the sensation of warmth is very consistently reported."

"Warmth in the emotional sense?" asked the Dalai Lama.

"Thoughts, feelings, and sensations seem to roll into a nondistinct singularity in this after-death state," said Joan. "There is also a feeling of painlessness, as if the after-death body does not experience the pain that this body experiences. In terms of visual experience, people have reported all different kinds of lights, manifestations, luminosities, auras, visual experiences of figures, and also the dark tunnel. Auditory experiences include everything from wind, to ringing, to buzzing sounds, even buzz saws, and heavenly choirs. The sense of smell has not been reported as being present, nor has the sense of taste. The sense that measures weight, motion, and position is also nonexistent. You're in a weightless state, you can't calibrate, and anything can happen to position."

"Do they feel touch?" Bob Livingstone asked. "Olfaction, taste, and touch all require immediate body contact with a chemical or an object."

His Holiness added, "This corresponds to Buddhist discussions of different types of heightened awareness. There is reference to clairaudience and clairvoyance, but there is no mention of heightened gustatory, olfactory, or tactile awareness. These three do not become elevated through meditation."

Pete Engel wanted to know if the visual and auditory senses stop functioning before taste and touch in the dying process. "The visual awareness goes first," replied the Dalai Lama, "and second is sound. The final stages of dissolution are smell, then taste, then touch last."

Joan said, "In my clinical experience, many of the people with whom I've worked in the final stages of life actually lose the sense of taste and of smell before they go through the death experience. This is also often true for old people. Usually smell diminishes first and then taste. A little remnant of taste remains, and then closer to death it disappears, sometimes several weeks before death. In the actual event of death, hearing goes last. Even people in coma can have an auditory response."

"This is something we need to reinvestigate in relation to the Buddhist teachings," His Holiness stated. "The stages of dissolution according to the Guhyasamāja Tantra need to be scrutinized once again in light of what we just heard. According to both the Sūtrayāna and the Vajrayāna systems, the final locus for heat within the body during the dying process is at the heart."

"Again, from my experience," said Joan, "the extremities grow very cold and the cold creeps up the body, and the last area of warmth perceived by individuals who are not yet dead is at this core."

His Holiness noted, "According to Buddhist tradition it disappears from the top down and the bottom up, lingering finally at the heart. It's considered better when the heat is withdrawn upwards from below. In the Sūtrayāna system, it's not so good if the heat disappears first from the top down."

"Physiologically the energy is carried by circulation and as the heart fails it's the distal parts of the circulation, the extremities and the head, that lose the energy first," clarified Pete.

His Holiness addressed the biomedical contingency in the room: "Isn't it true that the brain remains alive even for a minute or two after the heart stops? In that case, wouldn't the heat remain there for a while?"

Bob Livingstone pointed out that the brain accounts for two percent of the body's weight, but it consumes twenty percent of the oxygen and glucose and twenty percent of the circulation, so that it will remain comparatively warm. I pointed out, however, that if the brain is warm, we wouldn't feel it as such because there is no sensation in the brain. It's the scalp only that feels warmth in the head, so the two things are not contradictory.

"I was speaking precisely about the external sensations," the Dalai Lama said. "I'm not talking about a first-person account of what the dying persons feels, but rather an external account taking objective measurements of the dying person. From this third-person perspective, is it true that the last vestiges of heat are to be found at the heart?" Bob confirmed that heat remained longest in the chest, though not necessarily the heart. "I mentioned before that when we speak of the heart in this context we are not referring to the heart organ, but rather the center of the chest. So this accords with both Sūtrayāna and Vajrayāna accounts within Buddhism that the initial locus for consciousness at conception is at the heart; and similarly, the final locus of consciousness before death is once again the heart."

Bob smiled and said, "Yes, that would correspond. Except that for Western neuroscientists or physicians, heat doesn't have anything to do with consciousness."

His Holiness smiled back, "Of course, neurology doesn't talk about the clear light in the first place. In both of these cases, speaking of the original locus and the final locus, I'm speaking of this clear light, which is outside the parameters of neurology or neuroscience. Of course we've already mentioned that the gross consciousness dependent on the brain is extinguished."

Core Experiences

Joan steered the conversation back on track. "Two important cognitive transformations occur. We have already mentioned the transformation of the perception of time, which moves from chronological to timeless. The second transformation is that of mental clarity. Individuals in this state perceive themselves to be mentally more clear than in any other. A survivor of anaphylactic shock—an allergic reaction that can cause rapid death—says, 'I was fully aware of what was happening both physically and in the minds of those in the emergency room until I was revived, at which point I was very confused.' There are many accounts of this extremely precise mental clarity. When they come back to this life they're totally confused. In the spiritualist literature the new dead—those who have just died—are also described as being confused.

"Near-death experience is always characterized as a kind of journey. Inevitably, vehicles or modes of moving about are described. The experience also frequently has a certain narrative quality. My own mother died a clinical death and she described herself on a cruise ship that visited the ports of her past. She and my father had liked to take cruises. An Indian woman in a near-death account said she was riding an elephant. You can see the psychological and cultural conditioning. In other versions, the person is propelled at great velocities by an invisible force, or is spinning, or being sucked, or floating through a dark tunnel. Still other versions are something akin to birth imagery, being squeezed through the uterus or the vaginal canal into the after-death state. There are lots of other images which seem to be culturally conditioned, Your Holiness: walking through an immense culvert or pipe, spinning in dizzying spirals, descending into wells, or into caverns, following a luminous guide through a dark valley, and so on.

"All the accounts, however, move from darkness into light. The experience of the light is what researchers call the *core experience*, seemingly common across cultures. The light has been described as clear, white, orange, golden, or yellow, definitely of a different order than daylight. It's much more brilliant but soothing. One both sees it and simultaneously is caught in it. It seems to completely flood the mind. It also radiates mental qualities, primarily wisdom and compassion. In other words, there's no distinction between light and mind, as though the mind is actually a matrix for this light. In this moment of complete irradiation one seems to comprehend everything."

"I wonder whether people under anesthesia experience peace and pleasure simply due to the cessation of their experience before taking anesthesia, which was probably very uncomfortable," said His Holiness. "There may be a parallel here with the person approaching death. Before death, there's bound to be a lot of discomfort in the body and perhaps in the mind as well, and the very release from that discomfort would be experienced as peace and joy."

Joan replied, "The light is a visual experience, but it's combined with a sense of splendor, clarity, transparency, warmth, energy, a sense of boundless love and of embracing or all-encompassing knowledge. This is not only true of the accounts cross-culturally; this condition of light was described similarly in the medieval accounts.

"Then we arrive at what was called judgment in the medieval literature and what we call life review. It's been described by many individuals as a nonverbal colloquy between a being of light—not necessarily the big light—and the individual who's going through the experience. Very consistently, but not in all cases, there is the feeling of scrutinizing and evaluating one's own life. Frequently this happens as a kind of visual record. You may see your life in chronological order or in reverse order, or sometimes you see it all simultaneously in a moment. The vision of this life review is extremely vivid, as if the events are actually happening, even though time is transformed from the subject's perspective. Many contemporary subjects feel that the purpose of this review is to understand the progress one is making in one's life.

"I would like to read one account to give you a flavor. Notice that she speaks about herself in the third-person, with a great sense of objectification. This is a woman who was born in 1937. 'And into this great peace that I had become there came the life of Phyllis (her own name) parading past my view. Not as in a movie theater, but rather as a reliving. The reliving included not only the deeds committed by Phyllis since her birth in 1937 in Twin Falls, Idaho, but also a reliving of every thought ever thought, every word ever spoken, plus the effect of every thought, word, and deed upon everyone and anyone who had ever come within the sphere of her influence, whether she had actually known them or not, plus the effect of her every thought, word, and deed upon the weather, the air, the soil, plants and animals, the water, everything else within the creation we call earth and the space Phyllis once occupied. I had no idea a past life review could be like this. I never before realized that we were responsible and accountable for every single thing that we did. This was overwhelming. It was me judging me, not some heavenly Saint Peter, and my judgment was critical and stern. I was not satisfied with many things Phyllis had done and said or thought. There was a feeling of sadness and failure, yet a growing feeling of joy when the realization came that Phyllis had always done something. She did many things unworthy and negative but she did something, she tried, she tried. Much of what she did was constructive and positive. She learned and grew in her learning. This was satisfying. Phyllis was okay.'

"One characteristic both of the life review and of the light is the feeling of comprehensive knowledge. Another is that you often forget what it was that you knew when you're back. You come back here and you know you understood the meaning of life there and it's brought you incredible joy, but you can't quite remember what it was that you realized."

His Holiness responded, "This is an important phenomenon of forgetfulness in which a person remembers having known something very crucial, cannot recall it, yet feels affected by it. Is this a phenomenon common to all types of people who have near-death experiences, including those whose account can be verified? You gave an example of a person who vividly hears and follows the conversation and, when he or she returns, can tell you what has been going on elsewhere. Let's look at that in a class by itself, because it is very significant and clearly is not just a fabrication. Would such a person also experience this type of forgetfulness?"

Joan confirmed that, while forgetfulness does not happen in all cases, it happens often enough.

"The reason for my question is that if a person actually separates from the body and has various recollections of the experience, the recollections in that case would not be contingent on the brain. It would be quite separate. Then when the person comes back and awareness is once again contingent on the brain, it seems quite possible that memories from the noncontingent period would not transfer over to the subsequent contingent period. Whereas for a person who's having more of a dreamlike experience, all the mental processes would be contingent on the brain, in which case it would seem more likely that memories from the dreamlike period would be recollected afterwards."

Once more His Holiness was years ahead in designing experiments and testing evidence way beyond what seemed possible. Joan could only admit that it seemed indeed like an interesting direction for research, but that there was no available answer. She then continued her account about elements in the afterworld: utopian colors and blissful scenes including glowing lawns, flower-filled meadows, brilliant skies, shimmering bodies of water, rainbows, precious metals and jewels, and incredible architecture characterized by spaciousness and illuminated beauty.

Company and Well-Being

"Who else is there?" asked Joan. "One of the most common descriptions is a throng of angelic beings, particularly Christ and relatives. Finally there is the return. Most people don't want to come back. In fact, they get angry at the people who have resuscitated them. They're irritated, confused, and feel that they're being forced back into their body. They come back primarily because of the attachment they feel to their families. There is unfinished business. Sometimes they also have another, perhaps less important, reason to come back. They have to fulfill a mission related to the realization that they've had in this near-death state. It seems that these experiences have a very beneficial effect on people. The researchers report that people who have survived clinical death or near death have greater zest for life; their concern for material life is much diminished; they have greater self-confidence; and they feel a real sense of a purpose in life. They become spiritually enthusiastic, more interested in nature, and develop a tolerance and compassion toward others. Also, they have a reduced fear of death because they have become convinced that death is a great ride! They have a sense of relative invulnerability, so they feel very optimistic. They also feel as if they have a special destiny, perhaps a little pride that they died and were reborn.

"There are sometimes aftereffects of the near-death experience: a number of individuals have developed psychic powers: the ability to know in advance that something will happen, or telepathy, or out-of-body experiences. These experiences are considered, at least by the subjects, not only as a sanctuary but also as a high standard of mental functioning. These experiences are now entering Western psychology and are beginning to be considered not signs of pathology but in fact natural states of consciousness."

His Holiness asked whether any subjects have reported getting depressed, or have their lives degenerate following a near-death experience. Joan answered, "In fact, one of the consistent elements in all of these reports is their positive nature, whereas in the medieval tradition they're not so pleasant and sometimes even terrifying. It's peculiar to Western culture that the experiences have transformed from difficult and guilty experiences into pleasant, blissful, beautiful experiences.

"In summary, of the near-death population, 60 percent experience peace; 37 percent body separation or out-of-body experiences. Twenty-three percent entered the darkness, 16 percent saw the light, and 10 percent entered the light. The percentages follow the narrative. In other words, people abort the journey at different stages."

Some Materialistic Perspectives

Joan continued, "Finally the question arises: what does it mean to be dead? Harvard University developed four criteria: (1) unreceptivity and unresponsivity, in other words, the individual is not responding and is not receptive to external stimuli; (2) no movement and no breathing; (3) no reflexes; and (4) a flat electroencephalogram. The cardiologist Fred Schonmaker interviewed fifty-five patients who recovered from so-called brain death who had flat EEGs and met the Harvard criteria, and all of those had vivid memories of blissful visions."

I asked why the resuscitation would continue if somebody met the four criteria and was declared dead. Pete clarified a key point: "It was not mentioned that the Harvard criteria also require that a person be in this state for twelve hours. Up to that time they would continue to be respirated or supported, but they wouldn't be dead by the Harvard criteria. I'd like to know whether these people were in that state for twelve hours."

"I seriously doubt it," Joan said, "but it would be very interesting to look at that in more detail. In any case, it has become clear in our discussions that it's very difficult to pinpoint a particular moment of biological death. Cells deteriorate at different rates, physiological processes succumb at different velocities, and so a moment of death per se probably does not exist."

I asked His Holiness what would happen if the medical technologies of resuscitation so common in Western hospitals were applied to a Tibetan practitioner who was pronounced dead and experiencing the *bardo*. He answered, "If the person has gone as far as the clear light, but no farther, it's conceivable that by technological means it would be possible to bring this person back; but if the person has gone beyond the clear light into the *bardo*, it's very difficult. Once a person is determined to be clinically dead, how long thereafter is resuscitation possible? What's the limit?"

Pete answered, "It's actually about half an hour. It's certainly not the several days that practitioners can remain in the clear light, so I would find it difficult to believe that somebody could be resuscitated at the end of the clear light experience."

His Holiness admitted confusion. "I don't understand. When someone is pronounced clinically dead, does that mean the four criteria are true for twelve hours?"

"No," answered Pete. "I was describing brain death. *Clinical death* usually occurs when the heart stops beating and can't be restarted. In brain death, the body is still alive but its brain appears to be dead. Certain drugs that cause death can also cause this effect. The drugs decrease metabolism so they actually protect the brain from lack of oxygen and lack of glucose, enabling a person to survive longer in a state of death and still be resuscitated."

Joyce McDougall raised another skeptical voice: "Have you any idea of the proportion of people who experience near death and come back without these experiences?"

"No, but I think it's an important question to ask," added Joan. "I think it's interesting to look at some of the possible biological and psychological causes of near-death experiences in a materialistic perspective. Near-death experiences are caused when the nervous system is either overburdened or undernourished. So these experiences could occur under the influence of drugs or very stressful conditions that trigger hallucinations. In fact there are a number of drugs that can produce these experiences, from anesthesias to various opiate derivatives, to hallucinogens. Most of the researchers who work in this area are trying to posit a materialistic origin for these experiences. They say that the patients are suffering from severe imbalances of the metabolism, such as fever, insulin coma, exhaustion, trauma, infection, liver poisoning, or kidney failure. All of these can produce experiences resembling near-death experiences. Another is limbic lobe syndrome, a kind of seizure activity that produces depersonalization and involuntary memory recall."

Pete elaborated, "Epileptic seizures typically involve an area of the brain called the limbic system, which is believed to be related to motivations and to control the vegetative organs. During seizures the limbic functions take on bizarre forms and patients can have various experiences including depersonalization, out-of-body experiences, religious experiences, and perception of things changing shape and size."

Joan continued, "There are also certain chemicals produced in the body itself that produce very unusual experiences: endorphins, endopsychosins, and encephalins."

Seeing the Dalai Lama's questioning expression, I filled in some historical background on opiates. "Some fifteen or so years ago, neuroscientists discovered that the active ingredients of opium are produced inside the brain. These chemicals are called endogenous opiates, or endorphins. It was also discovered that when a person is under severe stress—such as facing death—these substances are released and cause analgesia, a loss of sensitivity to pain. Hence if somebody is confronting death, it is conceivable that he or she could in fact be having an opium experience. That's a possible reason why it would seem blissful and pleasant and very luminous."

Bob Livingstone elaborated, "Earlier, His Holiness was saying that if you have a hammer hitting your head, you feel better when it stops, a kind of a reciprocal pleasure. I'm sure that if you hit the head with a hammer, the brain stem would release a lot of endorphins which would not only protect you from the pain of the head blow but also produce a blissful feeling of serenity and calmness."

His Holiness asked whether the release of these chemicals explains the phenomenon of people who don't feel pain at the moment they are shot or knifed, and how long the effect remains. I explained that it could last for a long time, even hours.

Joan raised the extreme case of torture, and the possibility that at a certain point the agony transforms to ecstasy and reaches the brink of sexual ecstasy, including orgasm. This is also associated with the gallows: hanging actually produces penile engorgement and ejaculation. Women who are giving birth talk about orgasm at the moment of birth. Bob added that the release of endorphins can also be conditioned and excited in anticipation. "For example, a person who's going to play football and is expecting to be hurt by being tackled may have a high level of endorphins from the very beginning of the game. When he is tackled, he may have severe bruises, but it doesn't affect him. The same thing happens with soldiers going into battle. People who report the immediate effects of a wound, even a very severe one, often say they felt no pain."

Joan, however, qualified this discussion. "Your Holiness, the near-death researchers have found that the clearest, most vivid near-death experiences occur in individuals who have not been drugged, who have not suffered from sensory isolation, who aren't crazy or having a seizure. These pathologies, and the release of endorphins, are frequently followed by a sleepy, dreamy state which doesn't match the vivid state described by near-death survivors."

Possession and Epilepsy

Joan was concluding her presentation: "Here in this conference, we are exploring a realm where neither the neurological component, nor the phenomenological, ontological component of the experience is thrown out. We can look at them in a complementary and holistic fashion.

"From the point of view of someone whose eyes are open only during the day, the stars in the sky don't exist; but then in another state, called night, the stars suddenly appear. It's interesting to consider that the realm talked about in the *bardo* teachings may exist in a similar way. Western culture has often considered these experiences either pathological or unreal, and there has been an effort to reduce those experiences to physiological or biological conditions. We are now trying to really look at the physiological, biological, and neurological components in relation to the experiential component and not necessarily assume one origin for all. This brings me to a question that I'd like to pose: how do these experiences correlate with the experience that the *bardo* body has?"

"It's very difficult to relate near-death experiences as you have presented them to the experiences of a *bardo* being," His Holiness said, continuing to challenge the standard interpretation of near-death experiences in the West. "Not only that, but there is also the problem of the exact meaning given to the term *bardo state.* For instance, in Tibetan society we often hear of stories about spirits of dead people, but it's difficult to say whether or not they are in the *bardo* state.

"For instance, in the autobiography of the yogi Milarepa, a great Tibetan poet-saint who lived in the twelfth century, there is a story in which someone uses death rituals from the Bön tradition to summon the spirit of a dead person. Then with another ritual the spirit is prepared for its onward journey. At that time a member of the family actually saw a being with the same appearance as the person who passed away. Milarepa told the family member that in fact this was not really the spirit of the deceased relative, but rather some other spirit disguised in that form. The spirit of the deceased person had already taken rebirth in the form of an insect. Then Milarepa performed a transference of consciousness, *powa*, and it is said that a white rainbow was seen.

"I also know of a person who died and later began entering into, or possessing, different people. He was a monk, and a disciple of my junior tutor. A few days after he passed away, he possessed someone, placing this person in a trance. Speaking through the possessed person, he precisely described various things in his own room, and then he asked for certain things from my tutor. My tutor finally met him, through the possessed person. He gave him some advice on spiritual practice. Then after a few days, there were no more disturbances or possessions. But it's very difficult to say whether that spirit was in the *bardo* realm or somewhere else. According to popular Tibetan belief, it is said that these spirits have not been able to take rebirth, but it's very difficult to fit this belief within the Buddhist classification of six destinations or states of existence. Generally speaking, one spends a maximum of forty-nine days in the *bardo*, but there are accounts of these seemingly disembodied spirits remaining for as long as a year. That's why it's difficult to reconcile these accounts with Buddhist teachings."

"What happens to the person who is possessed?" asked Pete. "Does he come back, and does he have any recollection of this afterwards?"

"He will come back of course, but he won't have any recollections. If you were to look at such people who were possessed, you would in all likelihood think they were having an epileptic seizure. But from the Tibetan perspective, we would say they were affected by external influences, and that there's nothing wrong with their body. Rather, there is an external influence that is creating that apparent seizure."

"What is it about the behavior that would seem epileptic as opposed to just the behavior of the person who is possessing the body? Are there actual epileptic seizures?"

"The physical behavior of people while in such a trance seems to vary, so you really can't generalize. In some instances there seems to be a seizure and the body becomes very stiff. Especially immediately after coming out of the trance, the body is really quite stiff. Although this phenomenon of possession is very common in Tibetan society, from a strictly Buddhist point of view it's difficult to explain. The Tibetan word for 'possession' consists of two syllables meaning 'mindstream' and 'permeates.' I use the words very often, but I don't know—it presents a problem! It's not as if the mindstream of the possessed person is cast out and a new mindstream enters, nor is it that one separates from the other. It is something other than these two options. It may also happen that all the sensory and mental faculties of the possessed person are very active, yet at the same time there is an external controlling agency. The person doesn't have self-control, even though the senses are all intact. This is an area where we really need scientific research. I was wondering if you can see a particular structural framework in the brain of an epileptic person, because one way of determining this would be to do a scan on the mediums or on the epileptic person."

"It would be very interesting to do a scan on somebody who was possessed, or even just an EEG," Pete continued.

"I think it would be better to do it before possession, because you can identify an epileptic person even before a seizure. Also, there are different types of mediums. Some might be able to respond to questions put to them during trance, while seeming to retain a lot of their own personality, whereas others go through a very dramatic transformation as they enter the trance state. For instance, a few years ago, I met someone who mentioned to me that there was a medium nearby. I asked where the medium was, and it turned out he was the person I was speaking with. He was possessed right then! I thought this person was just a normal human being walking with me. Then there are other cases that are not at all like that."

Near-Death Experiences and Buddhist Teachings

"To respond more generally to the question about the nature of the near-death experience and Buddhist teachings, one would need to clarify first the three realms of desire, form, and formlessness. Or, in the Kālacakra system, there is a finer classification with either six or thirty-one realms. Basically we can say there are different worlds, different experiences; human life is just one of them. What we usually call spirits are some different form of life, beings who have a different body and mentality. Within the desire realm, and more specifically within the environment inhabited by human beings, there is quite a wide variety of other entities. Each has its own class or species name, you could almost say, and they're all cohabiting with us right here. Others, such as *devas*, may be elsewhere in the desire realm, but there's quite a wide variety of beings who are said to be amongst us in the world. Just like human beings, some different life forms are more compassionate, and some are more harmful. They have their own problems. Human interference in the realms of these spirit beings seems to be less common than interference from the opposite direction.

"If you make friends with some of these beings they can be useful as oracles—but as to the actual mechanism or process of how a medium enters into a trance and how possession takes place, I don't know! When dealing with oracles, of course, there is also the danger of fraud. But if you find a qualified and really reliable oracle, then sometimes this can be quite useful. Let's take the specific case of the Nechung Deva and oracle. The first account of this *deva* traces back to India, where it first manifested. A certain individual went up to Northern Amdo, in the high grasslands of northeastern Tibet, and the Nechung Deva followed him. Then in the eighth century, during the time of Padmasambhava, this same being came from Hor in northern Amdo, to Central Tibet, not far from Lhasa. This Nechung Deva is considered to be a god of the desire realm."

Joan persisted, "Suppose that one of your students or monks had a near-death experience similar to what we described today, and asked you what it meant. What would you say to them?"

"Since I cannot interpret such experiences right here now, how could I interpret them to this young monk! I don't know how to interpret these in general, let alone specifically from his perspective."

I ventured, "Is it not true that the moment of death—complete death with no coming back—is a terrifying experience and not a blissful, pleasant state for most people who are not trained?"

"There are two possible sorts of experience, and what occurs has everything to do with one's attitudes while alive, one's mental processes, and so forth. It is not simply the same for everybody. Even among ordinary beings there's a wide variety of experience. People who are compassionate throughout their lifetime or at least during their later years prior to death, would share similar experiences in the intermediate state. But the intermediate state would follow a different pattern for people who throughout their lifetimes have had a short temper, or are generally negative. There's also a difference related to the type of birth one will take in the next life.

"There is a phenomenon called 'returning from the dead' or *delok* in Tibetan. Recall the experience of the mother who instructed her daughter not to touch her body, and for one week her body was completely immobile, after which she woke up and described the places she had visited while her body was immobile. That could be seen as an example of the phenomenon of returning from the dead. In such a case has the connection between the gross body and the very subtle energy-mind been severed? It's very uncertain. We're not sure whether this person was even breathing, or whether there might have been subtle respiration during that period. We don't know, so there are various possibilities. If this were the special dream body, which moves around severed from the gross body, that would not imply that the very subtle energy-mind is severed. Or it might be that the very subtle energy-mind was severed from the gross body, went out, and then came back. The latter case is very problematic, because the woman had this experience, as far as we know, not as a result of very deep meditative practice, but rather as a result simply of her own karma combined with special circumstances that took place. It's very hard to believe that the very subtle energy-mind would have been totally dissociated from the gross body without a very deep meditative practice, but it's an open question.

"Generally speaking, if the relationship between the very subtle energy-mind and the gross body has been severed, this is irreversible. However, if one has very high realization, then it is possible in one's meditative practice to separate the very subtle energy-mind from the gross body and to bring it back. Two years ago, I met an old nun who was eighty years old. Unfortunately, she passed away last year. Actually I think she knew she would soon die. That day she offered me some important Tibetan books, but I told her she should keep them, so they stayed in her hut. She had lived in Dharamsala for almost thirty years, and in Tibet she lived for years in the Potala Palace. Many people, including some Westerners, would go to her for divination, because her divinations were extremely accurate. When I met her, she told me she had married and borne one son when she was about twenty-seven or twenty-eight years old. Then that son passed away, so she decided not to remain in the householders' way of life. She gave up all family affairs and traveled around. She came to a mountain behind Drepung Monastery, one of the most important Buddhist centers in Tibet. An old lama lived there, around eighty-eight years old, with about fifteen disciples. She stayed there a few months, and listened to some teachings. One day she saw two of his disciples actually flying from one side of the mountain to the other. There's no reason for her to lie, and it seems she was of sound mind. So perhaps, if that's true, this is one more thing to investigate.

"From a Buddhist viewpoint, our experience is based on the five inner and five outer elements. When your meditative experience gets deep enough to control the five inner elements, then there's the possibility of also controlling the five outer elements. Although space seems empty, once you develop the energies you can control it. When that happens, you can look right through solid things and walk in empty space as if it were solid. Just as there are subtle particles in all of us, so there are particles even in space."

"Your Holiness mentioned that some really advanced practitioners can sever the connection at the actual moment of death and then come back. What happens there?" I asked.

"Not only is that feat intentional but it's part of mainstream practice in Highest Yoga Tantra. It has to do with generating an illusory body, which is separated from the gross body. Wherever the illusory body goes, it is accompanied by the very subtle energy-mind. You separate these two from the gross body and then bring them back. That's part of the practice."

Near-Death Experiences and the Clear Light

Pete intervened with an important question for Westerners. "To a naive Westerner, the near-death experiences that Joan described, particularly the bright light, resembles the clear light that's experienced in the Tibetan practice of dying. Is the problem in making that association the fact that, to see the clear light, you need to be very experienced in this practice and that it doesn't just happen to anyone?"

His Holiness responded, "To begin with, it's not at all true that one needs to be an experienced yogi in order to experience the clear light. Everybody can have that encounter. It's quite feasible that the experiences of light that you've described are facsimiles of clear light. They have something in common with the clear light, for this reason: insofar as the subtle energies dissolve, then the experience of a subtle light becomes stronger and stronger. As those energies and mental faculties withdraw, there is a corresponding experience of inner light. So this could be something along the same lines as the actual clear light, but it would be very difficult to come back from the experience of the most subtle clear light occurring at the very moment of death, unless one had some kind of very deep spiritual experience, entailing control of the elements in the body."

I said, "So disruptions of normal gross brain phenomena, such as an experience of clear light caused by an anesthetic like ketamine that interrupts sensorimotor perceptions, would give rise to these facsimiles?"

"Perhaps, but the kind of disturbance that would lead to a facsimile of clear light would involve the withdrawal of the grosser levels of consciousness. Then these indications similar to the clear light could arise. Heart disease or the anesthetic you mention might be examples of this. But now here is a difference. In the case of sudden death, as in an automobile accident, this process of dissolution or withdrawal is extremely swift, and hence very difficult to ascertain. In a normal death, when the body is still in relatively good condition, then these processes occur gradually, and there is a greater chance to recognize them. Does such a normal, gradual process last a matter of minutes or hours? It's not the same in all cases; there's a wide variety. When ordinary people die, there's no control, and I think that may be due to a physical condition. Those who have gained some experience as a result of meditative training can have a certain degree of control of the duration or the pace of the dissolution."

His Holiness continued: "Listen carefully. Within the context of the Guhyasamāja Tantric system, included in Highest Yoga Tantra, there are explanations of body isolation, speech isolation, and mental isolation. Following mental isolation, the actual illusory body arises, and following that there is buddhahood. In cultivating body isolation, there are certain practices designed to withdraw the vital energies into the central channel. As this occurs, you experience the dissolution of the different elements—earth into water, water into fire, and fire into wind.

"There are also subjective signs that indicate these elemental dissolutions. However, when one first gains such experience, there is no real certainty as to the order in which the signs of these dissolutions occur. Sometimes one of the signs, for example the sign of smoke, will be much stronger; while at other times the sign of sparks might be much stronger. So it's not so certain. One's own experience of these different visions is something like a clear light. To draw an analogy, they would be like a lit-up movie screen when there's no film being projected. This is like the clear light experience, and within these the different visions arise: smoke-like and so forth. As I mentioned before, in the earlier stages of this practice, during the cultivation of body isolation, the order of these signs is not entirely certain. However, as one becomes more and more proficient in the practice, then the order becomes more and more definite. As in the case of body isolation, which is the first stage, there is a corresponding process during speech isolation and later during mental isolation, as you progress in the practice. The culmination of the third stage, mental isolation, is the accomplishment of the illusory body. The illusory body is accomplished only then, following the culmination of this entire process. In meditation, only at that time do you experience the actual withdrawal, or dissolution, of the different elements just as this occurs during the dying process. So this is not merely a facsimile of the process; it is exactly as in death; but it is now experienced through practice, with control. It is at that point that the knots at the heart center are loosened. There are six knots." He demonstrated with two hands facing inward, interlocking eight fingers. "At this point, the individual becomes capable of flying. The knots at the heart are unraveled upon the culmination of mental isolation, and at that time one experiences the metaphoric clear light. It's from that point that you can fly. The reason for that is from that point, by the power of your yogic practice you've gained complete control over the inner elements in the body."

Joan took up an earlier thread. "One of the fairly constant features in near-death experiences in the West is life review. There's a great emphasis in the West on biography and autobiography. In the East, this is perhaps less true, but it is rather curious that life review has not been part of your discussion. Does it ever occur?"

"It's possible," His Holiness replied. "I have gotten to know some people who, by yogic power of the type we've been discussing here, have been able to recollect previous lives. When I ask some of my friends a question, they usually respond as normal people. But sometimes, as a result of intensive practice in long, meditative retreats, some experience of deeper clarity occurs, and at that moment they recall their past life. Some of them have been able to recall twenty or thirty lifetimes, some even during the Buddha's time. That means the power of memory is increasing, so automatically the memory of this life would also increase.

"I was wondering about one thing which might be quite common. In the case of clarity of the mind, there is an enhanced power of recollection, and this might be explained in terms of people's aspirations or preoccupations. For instance, in the case of Buddhist practitioners, mundane preoccupations are not so strong, for they are chiefly concerned with spiritual matters. Whereas people who have less experience or who are not practitioners are chiefly preoccupied with affairs of this lifetime. When the clarity of consciousness is enhanced, consequently the power of recollection becomes very clear. For instance, according to the Sūtra system of Buddhist practice, when a practitioner is trying to cultivate the power of heightened awareness (Skt. *abhijña*; Tib. *mngon shes*), part of the practice is to repeatedly reflect on the very object that one desires to know. The case of people in these near-death experiences may be similar. This shows that mental energy can go only towards the object which you want."

I was curious about the cultural specificity of the visions. "These people in a near-death experience see Christ, or the saints, or their relatives—something that is familiar to them, obviously. When one reads the *bardo* descriptions one sees descriptions of peaceful and wrathful appearances wearing Indian clothes and adornments. Westerners say, 'Well, I don't think I'm going to see that. I've never seen anything like it.' Is it true that in fact everyone will have their own cultural projections on the experience?"

"That's probably the case. The whole presentation of deities within mandalas comes from India, and it thus draws upon Indian culture. It's very likely that a person from another culture would have different experiences. For this very reason a recent, extraordinary Tibetan scholar by the name Geshe Gendun Choephel said that since Buddhism came from India, the Sambhogakāya—the very subtle body of a buddha—is depicted as wearing a crown and ornaments of an Indian king. Whereas, he said, if Buddhism had originated in Tibet, then maybe the headdress would be of a Tibetan style. And if Buddhism had originated in China, then the Sambhogakāya might be depicted with a long beard. Generally speaking, if you want to say what the actual nature of the Sambhogakāya is, you have to say it is a form endowed with the greatest possible adornments, beauty, and perfections. It's an utterly perfect and absolutely sublime body. That's what you can say that is true. But as soon as that statement is made within a specific culture, then of course people look around them and try to imagine what this perfect body might look like; and they might think of the adornments of a king and so forth.

"Moreover, the Sambhogakāya is a Rūpakāya, a form body, and the very purpose of the Buddha manifesting a form body is for the sake of others. This being the case, the appearance assumed would be something appropriate for others, since it's intended for their service. It's not the case that there is some kind of intrinsic, autonomous form of this Sambhogakāya totally independent from those whom the Sambhogakāya is designed to help. Although this is a slightly divergent view within Tibetan Buddhist understanding of the form of the Sambhogakāya, many earlier Tibetan scholars have described the Sambhogakāya as a mere appearance for the sake of others. So it is witnessed from the other's perspective, and is only a relational appearance."

"What can one say about the *bardo* appearances which is not culturally tainted? Is there a fundamental structure on which these cultural manifestations are based?" I insisted.

"The very detailed descriptions of the visions of wrathful and peaceful deities that the person in the intermediate state experiences are very specific descriptions for practitioners following a particular Nyingma practice. Therefore, it is not the case that all Tibetans even will necessarily experience the same pattern of visions in the intermediate state."

The day and the conference were coming to an end, and Pete Engel aimed straight at a question that begged to be asked in a very personal way. "I have a final question. It's a personal question as a Westerner with a great fear of death, and as a scientist. I listen to all these discussions about the Buddhist concept of death and I find it very logical and very comforting, but I'm skeptical because I'm a scientist. Should I view what we heard from Joan of these near-death experiences in any way as a confirmation that should be encouraging for my belief in the Buddhist view of dying, or is it neutral and unrelated?"

His Holiness laughed heartily and said, "That you have to figure out for yourself! And continue to investigate. In some sense one can also look at the phenomenon of suicide as someone who's trying to get out of a difficult situation. To gain relief from the difficulty, they take their own life. All this is very much related to whether we have just one life or many lives. If there's only one life, then it's quite simple: if life really becomes unbearable, then you do what you like. These are really complicated matters. I think that because of the human mind, there are many different dispositions, and as a result the different religions and philosophies came to being. The important thing is the individual. It's very important that you find something appropriate and suitable for yourself as an individual. You should find something you can digest and make use of."

The only thing left to do was to bid each other farewell with bows, handshakes, and the exchange of presents, followed by taking pictures of a very memorable week.

Coda: Reflections on the Journey

Winding Down

ON FRIDAY EVENING the participants assembled at Kashmir Cottage, overlooking the hills of Dharamsala, for a quiet night of celebration. People gathered around to eat, to exchange impressions at the intellectual or personal level, and to sit on the veranda where the valley gave us a royal, golden sunset. Soon it would be time to pack our belongings. We would leave the following day for Delhi, and then onward to our respective homes. My companions and I seemed to share that high-spirited nostalgia one feels when embarking on a journey with fellow travelers who had been until recently mere names on a piece of paper.

On Saturday morning a few of us had an interview with the Dalai Lama. Our intention was to evaluate the conference and plan a future one. His Holiness was very encouraging, saying over and over how useful it was to have these in-depth, private, respectful dialogues. He strongly reaffirmed the need for another conference, Mind and Life V. For its topic we chose altruism and compassion as natural phenomena: their evolution, physiological basis, and social context. Dr. Richard Davidson was to be the scientific coordinator.

What We Learned

In reflecting over the entire course of events, I felt that we each had garnered a rich harvest of exchanges that would serve to sustain and enhance our own further research and practice. Significant advances were made in two major areas: near-death experiences (NDEs), and the distinctions concerning the various levels of subtle mind. The dialogue on lucid dreaming and the stages of sleep established by neuroscience also proved particularly illuminating.

Perhaps the most striking observation of the turn of events was the Dalai Lama's skeptical reception of the Western studies on NDEs. He seemed to be saying that these studies are misdirected. The trauma and shock that initiates these accounts and the ensuing events do not correspond to the sequential processes charted by centuries of observation of natural death. Furthermore, the qualitatively different experiential content also suggests, in his view, that NDEs are a process that is distinct from the stages of dissolution at death. His reflections on this issue are a strong caveat to many Westerners who have taken NDE accounts as predictive of what awaits them in their inevitable future. The findings of the conference strongly suggest that the Buddhist tradition can make an important contribution to current research in this area, and it seems that an entire reevaluation of the Western approach to NDEs is needed. I am not suggesting that the Tibetan traditions are correct in their views on NDEs; rather their extensive experience is massive evidence that Western research may have spawned some rather hasty interpretations.

His Holiness's exceptional teachings on death also allowed us to cover the issue of subtle mind in considerable depth. These teachings point to what might be called the "really hard problem" of consciousness for Westerners. These subtle levels of consciousness are by definition pre-individual—they are not person centered. As such, they appear to Western eyes as a form of dualism and are quickly dismissed. Yet the experience of the conference tells us that we cannot take these teachings so lightly, for with adequate interpretation, they can make a remarkable contribution to our understanding of the many levels of transition between what we call ordinary consciousness and death. It is important to note that these levels of subtle mind are not theoretical; instead, they are delineated rather precisely on the basis of actual experience, and they merit respectful attention by anybody who claims to rely on empirical science. This could become a remarkable avenue of research, one His Holiness would clearly welcome.

I think that the issue runs even deeper, for an understanding of these levels of subtle mind requires a sustained, disciplined, and well-informed meditation practice. In a sense, these phenomena are open only to those who are willing to carry out the experiments, as it were. That some form of special training is needed for firsthand experience of new realms of phenomena is not surprising. A musician also needs special training to have access to the experiences of, say, jazz improvisation. But in traditional science such phenomena remain hidden from view, since most scientists still avoid any disciplined study of their own experience, whether through meditation or other introspective methods. Fortunately, contemporary discourse on the science of consciousness increasingly relies on experiential evidence, and some scientists are beginning to be more flexible in their attitudes toward the first hand investigation of consciousness.

In the meantime, the epistemological gap between modern science and Buddhist teaching is a profound one. Only individuals who travel on both roads can provide bridges that avoid the pitfalls of reductionism. And building such bridges is a question of generations, not of a few discussions over a few conferences. As a part of that larger project, however, a clear chart of the really hard problems in the dialogue between science and Buddhism is surely invaluable. On this topic, Mind and Life IV produced the best chart so far.

In contrast to the divergent opinions on NDEs, Western and Buddhist theories on lucid dreaming confirmed each other. Many observations were made that were common to both traditions, and the physiological findings were easily correlated with theories of dream yoga. Notions about method were also quite compatible, although some of the advanced Tibetan yogas on illusory body seem to reach further than reported in the West. At the same time, Western technical knowledge may prove intriguing to a tradition that places so much emphasis on the development of lucidity.

Finally, the Dalai Lama appeared to think that the stages of sleep as described in neuroscience were a valuable addition to his tradition. The principal distinction between REM and non-REM and the transition between stages provided Buddhist observations with physiological validation, something that has always been an attractive possibility in these meetings.

Other participants might have chosen other items as highlights of the conference, but this is only further evidence of the richness of the event. The purpose of the conference was to continue to cultivate an opening of minds through a meaningful interchange between two traditions concerned with human life and mental experience. It clearly succeeded.

Return

During the trip back to Delhi through northern India, I saw the palpable tension of daily life. Hindu-Muslim conflicts flared, sparked by the Pakistan-India border conflict. To the west lay Afghanistan with its tragic civil wars; to the north, beyond the Hindu Kush, lay Central Asia and the lingering unrest in many of the newly independent republics; and on the far side of the snow-capped Himalayas the Chinese occupation of the ancient territory of Tibet continued. My return journey took me through Srinagar, once a jewel in the crown of the kings of Kashmir; now it seemed listless, drab, and polluted, stifled with makeshift buildings and overcrowded streets.

I could not help thinking that at the dawn of the twenty-first century the planetary social tissue is under as much strain as the earth itself. Human degradation, violence, and fundamentalist attitudes seem to overpower all social systems. What a stark contrast with the refined teachings of science and the Buddhist tradition of mind we had been considering. What a difference from the tolerance taught by the Tibetan community in Dharamsala and the spirit of respectful cross-cultural conversation we engaged in! Can the forces of growth and healing overcome those that threaten to tear the planet asunder? Whatever the answer, it will certainly require that we humans gain greater insight into our ability to transform our own experience.

Appendix

About the Mind and Life Institute

THE MIND AND LIFE DIALOGUES between His Holiness the Dalai Lama and Western scientists were brought to life through a collaboration between R. Adam Engle, a North American businessman, and Dr. Francisco J. Varela, a Chilean-born neuroscientist living and working in Paris. In 1984, both men independently had the initiative to create a series of cross-cultural meetings between His Holiness and Western scientists.

Engle, a Buddhist practitioner since 1974, had become aware of His Holiness's long-standing and keen interest in science and his desire to both deepen his understanding of Western science and share his understanding of Eastern contemplative science with Westerners. Varela, also a Buddhist practitioner since 1974, had met His Holiness at an international meeting in 1983, the Alpbach Symposia on Consciousness. Their communication was immediate. His Holiness was keenly interested in science but had little opportunity for discussion with brain scientists who had some experience with Tibetan Buddhism. This encounter led to a series of informal discussions over the next few years; through these conversations, His Holiness expressed the desire to have more extensive, planned time for mutual discussion and inquiry.

In the autumn of 1984, Engle and Michael Sautman met with His Holiness's youngest brother, Tendzen Choegyal (Ngari Rinpoche), in Los Angeles and presented a plan to create a week-long cross-cultural scientific meeting. Rinpoche graciously offered to take the matter up with His Holiness. Within days, Rinpoche reported that His Holiness very much wanted to participate in such a discussion, and authorized plans for the first meeting.

Meanwhile, Dr. Varela was moving forward with his own ideas. In the spring of 1985, a mutual friend, Dr. Joan Halifax, then the director of the Ojai Foundation, suggested that Engle, Sautman, and Varela could organize the first meeting collaboratively. The four gathered at the Ojai Foundation in October of 1985 and agreed to go forward. They decided to focus on the scientific disciplines that address mind and life, since these disciplines might provide the most fruitful interface with the Buddhist tradition. That insight provided the name of the project, and, in time, of the Mind and Life Institute itself.

It took two more years of work and communication with the Private Office of His Holiness before the first meeting was held in Dharamsala in October 1987. During this time, the organizers collaborated closely to find a useful structure for the meeting. Varela, acting as scientific coordinator, was primarily responsible for the scientific content of the meeting, issuing invitations to scientists, and editing a volume from transcripts of the meeting. Engle, acting as general coordinator, was responsible for fundraising, relations with His Holiness and his office, and all other aspects of the project. This division of responsibility between general and scientific coordinators has been part of the organizational strategy for all subsequent meetings. While Dr. Varela has not been the scientific coordinator for all of the meetings, he has remained a guiding force in the Mind and Life Institute, which was constituted in 1988 with Engle as its chairman.

A word is in order here concerning these conferences' unique character. The bridges that can mutually enrich traditional Buddhist thought and modern life science are notoriously difficult to build. Varela had a first taste of these difficulties while helping to establish a science program at Naropa Institute, a liberal arts institution created by Tibetan meditation master Chögyam Trungpa as a meeting ground between Western traditions and contemplative studies. In 1979 the program received a grant from the Sloan Foundation to organize what was probably the very first conference of its kind: "Comparative Approaches to Cognition: Western and Buddhist." Some twenty-five academics from prominent North American institutions convened. Their disciplines included mainstream philosophy, cognitive science (neurosciences, experimental psychology, linguistics, artificial intelligence), and, of course, Buddhist studies. The gathering's difficulties served as a hard lesson on the organizational care and finesse that a successful cross-cultural dialogue requires.

Thus in 1987, wishing to avoid some of the pitfalls encountered during the Naropa conference, several operating principles were adopted that have contributed significantly to the success of the Mind and Life series. These include choosing open-minded and competent scientists who have some familiarity with Buddhism; creating fully participatory meetings where His Holiness is briefed on general scientific background from a nonpartisan perspective before discussion is opened; employing gifted translators like Thupten Jinpa, Dr. Alan Wallace, and Dr. José Cabezón, who are comfortable with scientific vocabulary in both Tibetan and English; and finally, creating a private, protected space where relaxed and spontaneous discussion can proceed away from the Western media's watchful eye.

The first Mind and Life Conference took place in October of 1987 in Dharamsala. The conference focused on the basic groundwork of modern cognitive science, the most natural starting point for a dialogue between the Buddhist tradition and modern science. The curriculum for the first conference introduced broad themes from cognitive science, including scientific method, neurobiology, cognitive psychology, artificial intelligence, brain development, and evolution. In attendance were Jeremy Hayward (physics and philosophy of science); Robert Livingstone (neuroscience and medicine); Eleanor Rosch (cognitive science); and Newcomb Greenleaf (computer science). At our concluding session, the Dalai Lama asked us to continue the dialogue with biennial conferences. Mind and Life I was published as *Gentle Bridges: Conversations with the Dalai Lama on the Sciences of Mind*, edited by Jeremy Hayward and Francisco Varela (Boston: Shambhala Publications, 1992). The volume has been translated into French, Spanish, German, Japanese, and Chinese.

Mind and Life II took place in October 1989 in Newport, California, with Robert Livingstone as the scientific coordinator. The two-day conference focused on neuroscience. Invited were Patricia S. Churchland (philosophy of science); J. Allan Hobson (sleep and dreams); Larry Squire (memory); Antonio Damasio (neuroscience); Robert Livingstone (neuroscience); and Lewis Judd (mental health).

Mind and Life III was held in Dharamsala in 1990. Daniel Goleman (psychology) served as the scientific coordinator. He chose to focus on the relationship between emotions and health. Participants included Dan Brown (experimental psychology); Jon Kabat-Zinn (medicine); Clifford Saron (neuroscience); and Lee Yearly (philosophy). Mind and Life III is available as *Healing Emotions: Conversations with the Dalai Lama on Mindfulness, Emotions, and Health*, edited by Daniel Goleman (Boston: Shambhala Publications, 1997).

During Mind and Life III a new mode of exploration emerged: participants initiated a research project to investigate the neurobiological effects of meditation on long-term meditators. To facilitate such research, a Mind and Life network was created to connect other scientists interested in both Eastern contemplative experience and Western science. With seed money from the Hershey Family Foundation, the Mind and Life Institute was born. The Fetzer Institute funded two years of network expenses and the initial stages of the research project. Along with the publication of many of our findings, research continues on various topics, such as attention and emotional response.

Following Mind and Life IV (the subject of the present volume), Mind and Life V convened in Dharamsala in October 1994. This conference focused upon the exploration of altruism, ethics, and compassion as natural phenomena; our intent was to examine and discuss the evolution, physiological basis, and social context of these closely linked issues. Dr. Richard Davidson served as the scientific coordinator. He is currently editing the proceeds as *Science and Compassion: Dialogues with the Dalai Lama*. In attendance were

Richard Davidson (cognitive neuroscience); Anne Harrington (history and philosophy of science); Robert Frank (altruism in economics); Nancy Eisenberg (child development); Ervin Staub (psychology and group behavior); and Elliott Sober (philosophy).

Mind and Life VI is scheduled for October 1997. The participants will once again meet in Dharamsala. For the first time, the topic will shift from the biological sciences to physics and cosmology. Arthur Zajonc (Amherst College) has agreed to serve as the scientific coordinator, and Adam Engle will reprise his original role as the general coordinator.

ACKNOWLEDGMENTS

Over the years, the Mind and Life conferences have been supported by the generosity of several individuals and organizations. Barry and Connie Hershey of the Hershey Family Foundation have been our most loyal and steadfast patrons since 1990. Not only has their generous support guaranteed the continuity of the conferences, but it has breathed life into the Mind and Life Institute itself. Over the years the conferences have also received generous financial support from the Fetzer Institute; The Nathan Cummings Foundation; Mr. Branco Weiss; Adam Engle; Michael Sautman; Mr. and Mrs. R. Thomas Northcote; Ms. Christine Austin; and Mr. Dennis Perlman. On behalf of His Holiness the Dalai Lama and all the other participants, we humbly thank all of these individuals and organizations. Your generosity has had a profound impact on the lives of many people.

We would also like to thank a number of people for their assistance in making the work of the Institute itself a success. Many of these people have assisted the Institute since its inception. We thank and acknowledge His Holiness the Dalai Lama; Tenzin Geyche Tethong and the other wonderful people of the Private Office of His Holiness; Ngari Rinpoche and Rinchen Khandro, together with the staff of Kashmir Cottage; all the scientists, scientific coordinators, and interpreters; Maazda Travel in the United States and Middle Path Travel in India; Pier Luigi Luisi; Elaine Jackson; Clifford Saron;

Zara Houshmand; Alan Kelly; Peter Jepson; Pat Aiello; Thubten Chodron; Laure Chattel; Shambhala Publications; and Wisdom Publications.

The Mind and Life Institute was created in 1990 as a 501 (c) 3 public charity to support the Mind and Life dialogues and to promote cross-cultural scientific research and understanding. We can be reached at P.O. Box 94, Boulder Creek, CA 95006. Telephone: (408) 338-2123; fax: (408) 338-3666; E-mail: aengle@engle.com.

—Adam Engle

Notes

1. Mind and Life I was published as J. Hayward and F. J. Varela, eds., *Gentle Bridges: Conversations with the Dalai Lama on the Sciences of Mind* (Boston: Shambhala Publications, 1992).

2. Mind and Life III was published as D. Goleman, ed., *Healing Emotions: Conversations with the Dalai Lama on Mindfulness, Emotions, and Health* (Boston: Shambhala Publications, 1997).

3. C. Taylor, *Sources of the Self: The Making of the Modern Identity* (Cambridge, Mass.: Harvard University Press, 1990).

4. The interested reader may consult, for instance, J. A. Hobson, *The Sleeping Brain* (New York: Penguin Editions, 1991).

5. See, for instance, K. M. Colby and R. J. Stoller, *Psychoanalysis and Cognitive Science* (Hillsdale, N.J.: Analytic Press, 1988).

6. J. McDougall, *Theaters of the Mind* (New York: Basic Books, 1985); *Theaters of the Body* (New York: W. W. Norton, 1990).

7. For a very informative biography of Freud, see Peter Gay, *Freud: A Life for Our Times* (New York: Anchor/Doubleday, 1988).

8. G. Roheim, *The Gates of the Dream* (New York: Macmillan, 1965).

9. See, for example, D. W. Winnicott, *Playing and Reality* (New York: Basic Books, 1971).

10. J. Gackenbach and S. LaBerge, eds., *Conscious Mind, Sleeping Brain* (New York: Plenum Press, 1988).

11. For a general introduction, including historical sources, see S. La Berge, *Lucid Dreaming* (Los Angeles: Tharcher, 1985).

12. See S. LaBerge, L. Levitan, and W. C. Dement, "Psychophysiological Correlates of the Initiation of Lucid Dreaming," *Sleep Research* 10 (1986): 149.

13. See S. LaBerge, "Psychophysiology of Lucid Dreaming," in Gackenbach and

LaBerge, eds.,*Conscious Mind, Sleeping Brain*, 135-52.

14. Among the remaining forty-seven mental factors are five omnipresent mental factors, five object-ascertaining mental factors, eleven wholesome mental factors, six primary mental afflictions, and twenty secondary mental afflictions.

15. Philippe Ariès, *Histoire de la Mort en Occident* (Paris: Seuil, 1974).

16. See D. Griffin, *Animal Minds* (Chicago: University of Chicago Press, 1989).

17. C. Trungpa, *Transcending Madness: Bardo and the Six Realms* (Boston: Shambhala Publications, 1993).

18. See Hayward and Varela, eds., *Gentle Bridges*.

19. J. E. Ahlskog, "Cerebral transplantation for Parkinson's Disease: Current Progress and Future Prospects," *Mayo Clinic Proceedings* 68 (1993): 578-91.

20. J. Engel, Jr., *Seizures and Epilepsy* (Philadelphia: F. A. Davis, 1989), 536.

21. See F. Fremantle and C. Trungpa, trans., *The Tibetan Book of the Dead* (Boston: Shambhala Dragon Editions, 1992).

22. See R. Moody, *Life after Life* (Atlanta: Mockingbird, 1975); K. Ring, *Heading Towards Omega* (New York: Quill Morrow, 1984); M. Sabom, *Recollections of Death* (New York: Harper & Row, 1982).

Glossary

Abhidharma (Tib. *mngon chos*). Detailed philosophical investigations into the mind and mental functions, and the effects of various positive and negative mental states. The discipline of Abhidharma is said to have been started by the Buddha, and it continues to the present time.

afflictive obstructions (Skt. *kleśāvaraṇa;* Tib. *nyon mongs kyi sgrib pa*). The negative mental states and emotions which obscure the nature of reality and fuel the process of rebirth in cyclic existence, or *saṃsāra.* There are various enumerations of the negative mental states that comprise this obstruction, but all of them can be subsumed in the categories of the three poisons: attachment, anger, and ignorance.

ālayavijñāna (Tib. *kun gzhi nams shes*). Often translated as the "storehouse consciousness" or the "foundation consciousness," it is the most subtle of the eight forms of consciousness enumerated in the Yogācāra system. The foundation consciousness is the place where the propensities or imprints created by actions under the influence of afflictive obstructions are "stored" until the conditions for their manifestation are present.

Ārya Asaṅga. Indian Buddhist philosopher (c. fourth century C.E.) who helped to establish the Yogācāra system through his extensive writings on Abhidharma and other topics.

association cortex. Regions of the cerebral cortex that are connected to, and therefore integrate, diverse sensory (i.e., visual, auditory, tactile, etc.) or motor information. These regions are considered higher-order in that they allow for reflective, purposeful action.

asura. A member of the *asura* realm, one of the six realms of cyclic existence (*saṃsāra*) in Buddhist cosmology. Often referred to as *jealous gods*, *asuras* are born in a celestial realm through the force of both powerful positive *karma* and negative *karma.* Though blessed with wealth, intelligence, and long life, they are jealous of the superior wealth of the *devas*, and so are constantly at war with them.

awakening (Skt. *bodhi*; Tib. *byang chub*). The ultimate goal of the Buddhist path. One is considered to be awakened, or enlightened, when one has purified both the afflictive obstruction and the obstruction to knowledge, and when one has achieved the various qualities of a buddha. A person who has reached awakening is called a buddha, or Awakened One.

axon. The output fiber of the neuron that transmits information in the form of electrical impulses and chemical signals to other, usually distant, neurons.

bakchak. See *imprints.*

bardo (Skt. *antarābhava*). Refers generally to the *intermediate state* between death and rebirth, where the mindstream wanders in the form of a "mental body" while seeking a new embodiment. The *bardo* is considered to be an important opportunity for tantric practice, for it is at the point of transition from death into the *bardo* that the *clear light* nature of consciousness becomes manifest. Likewise, during the *bardo*, the mind experiences numerous appearances, said to be in the form of peaceful and wrathful deities. If the practitioner can realize these appearances as the nature of mind itself, he or she can attain liberation. See also *Bardo Thödol, clear light.*

Bardo Thödol (Tib. *bar do thos sgrol*). A famous Tibetan manual for the dying known in English as *The Tibetan Book of the Dead*; literally, the title means *Liberation through Hearing in the Intermediate State.* Reading this book aloud in the presence of someone who is dying or has already died is thought to help the dead person to recognize the phenomenal appearances of the *bardo* as the nature of mind itself and thus attain awakening.

basic clear light. The conceptually unstructured, primordial state of awareness that defies all logical categories, including existence, nonexistence, both existence and nonexistence, and neither existence nor nonexistence. See also *clear light.*

Bön. The indigenous religious tradition of Tibet. Although Bön possesses a history and mythology distinct from Buddhism, it has assimilated many of the philosophical views and meditative practices of the Tibetan Buddhist tradition at large.

brain stem. A general term for three neuroanatomical structures: the medulla, pons, and midbrain. The brainstem extends from the base of the brain, anterior to the spinal cord, to the center of the brain. It processes sensation from the skin and joints of the head, neck, and face, as well as specialized senses such as hearing, taste, and balance. The neural networks in these regions are also implicated in regulating the various brain "states" such as waking consciousness and the different phases of sleep.

central channel (Skt. *avadhūtī*; Tib. *rtsa dbu ma*). The primary "nerve" or "vein" of the subtle nervous system according to Tantric Buddhist physiology. By causing psychic energy or wind (Skt. *prāṇa*; Tib. *rlung*) to circulate in the central channel through yogic techniques, it is possible to recognize the fundamental clear light.

cerebral cortex. The thin, convoluted surface of the cerebral hemispheres that consists of the cell bodies of neurons (e.g., grey matter). It is divided into four main areas: frontal, parietal, temporal, and occipital.

cerebrospinal fluid. The fluid found within the four cavities, or ventricles, of the brain. The ventricles are connected to each other, allowing for the circulation and renewal of fluid. Cerebrospinal fluid serves a number of purposes, among them protecting the brain from forces (such as gravity) that may distort it and regulating the extracellular environment.

cerebrum. The two cerebral hemispheres of the brain, apart from the cerebellum and brain stem.

chakra (Skt. *cakra*; Tib. *'khor lo*). Literally, a "wheel" or junction of energy channels along the central channel; major chakras are located in the crown of the head, the throat, the heart, and the genital region.

chosen deity (Skt. *iṣṭadevatā*; Tib. *yidam*). In Buddhist tantra, the deity that forms the focus of an individual's tantric practice. The Buddhist tantras depict a vast pantheon of such deities, each with different attributes that are intended to best suit the propensities of particular practitioners. Meditation upon one's chosen deity is generally used to prepare the practitioner for the actual manipulation of the *vital energy* so as to facilitate the realization of the *clear light* nature of the mind.

circadian rhythm. A biological activity that occurs in approximately 24-hour periods or cycles, the most well-known of which is human sleep.

clear light (Skt. *prabhāsvara*; Tib. *'od gsal*). The subtle appearance that occurs when the vital energies have become absorbed into the central channel. The vital energies become absorbed in this way at several junctures, most notably at sleep and death or in tantric meditation. As the energies become absorbed into the central channel, the mind goes through the *eightfold stages of dissolution* that include a series of appearances culminating with the clear light itself. The experience of the clear light, which is said to be like a "clear, cloudless autumn sky just before dawn," represents the mind at its subtlest, and awareness of it is called the *natural clear light.* When the practitioner maintains awareness of it, she has realized the fundamental nature of the mind itself, in that the clear light is the subtle basis for all other mental content. Although extremely subtle, the *clear light of sleep* is not as subtle as the *clear light of death* because the vital energies do not become completely absorbed into the central channel upon falling asleep. At death, however, the energies do become completely absorbed, and for this reason, the clear light that appears at death is called the *basic clear light,* or *primordial clear light,* for it is the mind in its subtlest and most fundamental state.

clear light of death. See *clear light.*

clear light of sleep. See *clear light.*

consciousness (neuroscientific definition). A consensual neuroscientific definition is not available; however, the term has been used in association with the following: reflective awareness, attentional selection of environmental stimuli and monitoring of behavior, certain levels of wakefulness, synthesis of cognitive processing, etc.

consciousness (Buddhist definition). Most Buddhist philosophers define *consciousness* as "luminous awareness." The term *luminous* (Skt. *prabhāsvara*; Tib. *gsal ba*) refers to the capacity to "illuminate" or present objects. At the same time, consciousness is *luminous* because it is "clear," since it is like an open space which holds content but has no intrinsic content in and of itself. Finally, the luminosity of the mind refers to its basic nature, the *clear light.* While being luminous, consciousness is also "awareness" (Skt. *jñāna;* Tib. *rig pa*) because it also knows or apprehends the objects appearing to it. Hence, when one sees the color blue, the luminosity of consciousness accounts for the appearance of blue in the mind, and the awareness aspect of consciousness is what allows one to apprehend and then manipulate that appearance with other mental functions, such as conceptuality or memory.

cooperative condition (Skt. *sahakārīpratyaya*; Tib. *lhan cig byed rkyen*). A condition which must be present in order for a particular *substantial cause* to bring about its effect; for example, a seed must have the cooperative conditions of soil, moisture, and light to form a green sprout.

cyclic existence. See *saṃsāra.*

deity yoga. The practice of visualizing oneself as a buddha-deity in Tantric Buddhist practice. By imagining oneself in the state of the result (buddhahood), it is said to be possible to cultivate the necessary causes for full enlightenment in a single lifetime. See also *chosen deity; Vajrayāna.*

delok (Tib. *'das log*). An extreme form of "near-death" experience. Following an illness or accident, a person may lie in a state of suspended animation for several days, witnessing the effects of *karma* which must be suffered in the intermediate state and future lifetimes, and then return to life.

desire realm (Skt. *kāmadhātu*; Tib. *'dod khams*). In Buddhist cosmology, one of the three dimensions of existence (along with the form and formless realms). It is considered to be the least refined, since the bodies and minds of the beings in that realm are coarse. It is called the *desire realm* in part because the beings within it are motivated primarily by desire. Hell realm beings, hungry ghosts, animals, humans, jealous gods, and some *devas* are considered to be part of this realm.

deva (Tib. *lha*). In Sanskrit, *deva* refers to a "god" or celestial being. Some *devas* occupy the highest level within the desire realm of cyclic existence (*saṃsāra*), while others abide in the form and formless realms. *Devas* have extremely long lifespans, and they enjoy sensual and meditative pleasures; for these reasons, they generally lack the renunciation necessary for the path to awakening.

Dharmakāya. See *three kāyas.*

Diamond Vehicle. See *Vajrayāna.*

dream body. The apparent physical form one has in the dream state. In the yogic practice of the stage of completion within the Highest Yoga Tantra, the dream body is cultivated as a simulacrum of the illusory body.

dream yoga. A practice, similar to what is called in the West *lucid dreaming,* in which one cultivates awareness of the nature of dreams in order to use the dream state for spiritual practice.

drongjuk (Tib. *grong 'jug*). A yogic practice whereby a yogic practitioner transfers his or her consciousness to a dead body and revives it. Traditionally it is said that this practice was brought to Tibet by Marpa the Translator (1012-1097), but was lost when Marpa's son died an untimely death before he passed on the instructions for it.

Dzogchen (Tib. *rdzogs pa chen po*). Usually translated as *Great Perfection,* Dzogchen is the highest system of tantric meditation according to the Nyingma school of Tibetan Buddhism. In this system, a yogi or yogini cultivates direct, unmodified recognition of the Dharmakāya in dependence upon the introduction to the nature of mind received from a qualified lama, in conjunction with the lama's personal guidance.

effulgent pristine awareness (Tib. *rtsal gyi rig pa*). The form of pristine awareness that can become manifest even in waking consciousness without the practitioner being actually absorbed in meditation. Sometimes said to occur "between thoughts," it is similar to the other forms of pristine awareness in that it is the basic, primordial nature of consciousness itself. It is called "effulgent" in that it manifests as cognitive manifestations or appearances in consciousness. As such, it is the basis for all mental content.

eightfold process of dying/dissolution. According to Highest Yoga Tantra, as the five forms of vital energy dissolve into the central channel at death, a series of appearances occur in the mind of the dying person: mirage, wisps of smoke, fireflies, glowing lamp, white appearance, red "increase," black "attainment," and finally, the clear light of death itself.

electroencephalogram (EEG). An apparatus that records electrical activity produced by underlying brain processes. It involves the attachment of sensors, or electrodes, to the outer surface of the head.

electromyogram. Electrical activity recorded from a muscle or a muscle group.

emptiness (Skt. *śūnyatā*; Tib. *stong pa nyid*). The Buddhist philosophical position that things (*dharma*) are *empty* of any unchanging, intrinsic essence or existence. Although things are *ultimately* empty, they nonetheless can be said to exist *conventionally*, or in dependence on causes and conditions. See also *identitylessness.*

evident phenomena (Skt. *dṛśyadharma*; Tib. *mthong rung gi chos*). One of three classifications of things (*dharma*) in the Indian and Tibetan Buddhist epistemological tradition. Evident phenomena are things which are directly perceptible by means of any of the five senses and the mind.

extremely obscure phenomena (Skt. *ativiprakṛṣṭadharma;* Tib. *shing tu lkog gyur gyi chos*). One of three classifications of things (*dharma*) in the Indian and Tibetan Buddhist epistemological tradition. Extremely obscure phenomena are things whose existence one can neither directly perceive nor infer. Certain things may be extremely obscure for some persons, while not so to others. Hence, for ordinary persons, some aspects of the workings of *karma*, for example, are extremely obscure, but such *karmic* processes are evident to buddhas.

extremely remote phenomena. See *extremely obscure phenomena.*

five aggregates (Skt. *pañcaskandha*; Tib. *phung po lnga*). According to Buddhist philosophical thought, the mind-body system is divided into five aggregates, or constituents. The five are form, feeling, perception, volition, and consciousness. In the view of many Buddhist philosophers, these five aggregates form the basis upon which an individual's sense of "self" and personal identity is designated.

five inner and five outer elements. The five elements are earth, water, fire, air, and space. When they refer to the elements of which the body is composed, they are called the *five inner elements.* When they refer to the elements of which the outer universe is composed, they are called the *five outer elements.*

form realm (Skt. *rūpadhātu*; Tib. *gzugs khams*). In Buddhist cosmology, one of the three dimensions of existence (along with the desire and formless realms). The form realm beings are *devas*, but they are different from the *devas* of the desire realm in that they have eliminated all desires other than the desire for visible, audible, and tactile sense objects. One is born in the form realm as a result of meditating upon and perfecting one of four concentrations (*dhyāna*).

formless realm (Skt. *ārūpyadhātu*; Tib. *gzugs med khams*). In Buddhist cosmology, one of the three dimensions of existence (along with the desire and form realms). The beings in the formless realm have neither desire nor any sort of physical form. One is born in this realm as a result of meditating upon and perfecting one of four meditative concentrations (*samādhi*).

foundation of all. See *ālayavijñāna.*

foundation consciousness. See *ālayavijñāna.*

Four Noble Truths. The Four Noble Truths are (1) suffering exists; (2) the source of suffering is attachment; (3) there is a cessation of suffering; and (4) there exists a path leading to that cessation. All Buddhist traditions agree that these four principles lie at the heart of the Buddha's spiritual message. There are, within this formula of the Four Noble Truths, two sets grouped according to cause and effect. The first is associated with cyclic existence: truth of the source (the cause) and truth of suffering (the effect). The second set is related to liberation from cyclic existence: truth of the path (the cause) and the truth of cessation (the effect, liberation itself). In brief, the teaching on the Four Noble Truths outlines the Buddhist understanding of the nature of both *saṃsāra* and *nirvāṇa.*

frontal lobe. The anterior portion of the cerebral cortex comprising close to one third of the total cortical mass. Its operations are diverse and not easily definable. They include: the higher-order synthesis of information from virtually all other regions of the brain, executive processes, reason, judgment, interpretation of environmental contexts leading to appropriate social interaction, motor preparation/initiation/action, etc.

fundamental clear light. See clear light.

Gelug order. The most recently established of the Tibetan Buddhist schools of the New Translation lineage, the Gelug tradition was founded by the great scholar and yogi, Tsongkhapa (1357-1419). This school is particularly well known for its attention to philosophical study and debate.

geshe. The Tibetan term *geshe* literally means "spiritual friend." Currently, this title is generally conferred in the Gelug order on those who have successfully completed many years of monastic education and have thus attained a high degree of doctrinal learning.

Great Perfection. See *Dzogchen.*

gross body/gross consciousness. The ordinary state of body and mind conditioned by *karma* and negative mental states; they are the basis which is purified into the subtle or illusory body and pristine awareness which together are the essence of awakening.

Guhyasamāja system. One of the most important systems of Highest Yoga Tantra in the traditions of New Translations. The Guhyasamāja Tantra is the *locus classicus* for the practices of the stage of completion as codified by Nāgārjuna in his *Five Stages* (*Pañcakrama*).

heart-center. May refer to the heart chakra in general, or specifically to the center of the heart chakra, where the subtle wind and vital energy, which are the basis for stage of completion in Highest Yoga Tantra, are said to reside as a small sphere.

Highest Yoga Tantra (Skt. *anuttarayogatantra*; Tib. *bla na med pa'i rnyal 'byor*). The highest system of tantric theory and practice according to the New Translation traditions. It is distinguished, among other ways, by its advanced techniques for the manipulation of the vital energies, and by making full enlightenment possible in a single lifetime. See also *stage of completion, stage of generation.*

hippocampus. A deep-lying structure in the temporal lobes of the cerebral hemispheres. It is involved with aspects of memory, most notably the consolidation and storage of information that is consciously apprehended.

identitylessness (Skt. *niḥsvabhāvatā, anātmatā*; Tib. *rang bzhin med pa, bdag med pa*). The doctrine of *identitylessness* or *selflessness* is a key philosophical concept in Buddhism. In brief, it relates to the Buddha's insight that the state of unenlightened, conditioned existence is rooted in a false belief in the existence of a permanent, enduring self, essence, or identity. It is an insight into the absence of such a self that opens the door to liberation from the suffering of cyclic existence, or *saṃsāra.*

illusory body (Skt. *māyākāya*; Tib. *sgyu lus*). In Buddhist tantra, an illusory body is attained by an advanced practitioner in meditation or at the time of entering the intermediate state. Taking the extremely subtle wind, or vital energy, as the substantial cause and the mind as the cooperative condition, she or he arises in the form of a pure or impure illusory body. The Buddha's Sambhogakāya is, in effect, a pure illusory body. See *three kāyas.*

imprints (Skt. *vāsanā*; Tib. *bag chags*). Also referred to as *bakchak* and as latent propensities, the imprints are the habitual tendencies created by *karma,* which are said in the Yogācāra system to reside in the foundation consciousness. When these imprints encounter the necessary conditions, they will manifest as effects of the original *karma.*

increase of appearance. See *eightfold process of dying/dissolution.*

intermediate state. See *bardo.*

isolation: body, speech, and mental. The first three stages of the five stages of the stage of completion in the Guhyasamāja system. After acheiving these three, one goes on to achieve the *clear light* and the *illusory body.* The final unification of these latter two is equivalent to buddhahood.

jealous god. See *asura.*

Kālacakra Tantra. By some reckonings, the highest tantric system of the Highest Yoga Tantra. In addition to being the basis for important meditative practices, the *Kālacakra Tantra* is also a major source for mathematics, astrology, and prophecy as understood in Tibetan Buddhism.

karma (Tib. *las*). The Sanskrit term *karma* refers to actions and their imprints within the mindstream. The actions involved may be physical, vocal, or mental. In its general usage, *karma* refers to the entire process of causal action and the resultant effects.

lama (Skt. *guru;* Tib. *bla ma*). A qualified spiritual guide or any revered teacher is referred to in Sanskrit as *guru,* in Tibetan as *lama.*

Lamarckian evolution. Evolutionary theory propounded by French naturalist Jean de Lamarck. It is based on the assumption that species develop through the efforts of an organism to adapt itself to new conditions, and by subsequent transmission to descendants of the changes thus produced.

latent propensities. See *imprints.*

lateralization. The localization of cognitive functions in either the right or left hemisphere. For instance, in most right-handed people aspects of language, such as phonemic and syntactic processing and speech output, are understood to be mediated by the left hemisphere, whereas other language processes, such as interpretation of intonation and metaphor, are considered right hemispheric.

Madhyamaka. One of the four major philosophical systems of ancient Indian Buddhist thought, the Madhyamaka, or Middle Way, was founded by Nāgārjuna. It is best known for its articulation of the ultimate identitylessness of all persons and things and for steering a "middle way" between essentialism and nihilism through its postulation of the two truths (conventional and ultimate). There are two major schools within Madhyamaka, the Prāsaṅgika and the Svātantrika.

magnetic field recording (MEG). Measurement of the magnetic components associated with the electrical field of an EEG.

Mahāyāna (Tib. *theg pa chen po*). Literally, the "Great Vehicle." The Mahāyāna is one of the two main traditions that emerged within Buddhism in ancient India, the other being classified by the Mahāyāna as the Hīnayāna, or "Small Vehicle." Associated with the Buddhist traditions of Tibet, China, Japan, Korea, and Vietnam, a key characteristic of the Mahāyāna is its insistence on an altruistic and compassionate sense of universal responsibility for the welfare of all sentient beings as essential for attaining awakening.

Maitreya. The coming buddha and the embodiment of the loving kindness of all the buddhas. His name literally means "loving one." There is also a bodhisattva Maitreya as well as a historical person by the same name who is the author of several important Mahāyāna philosophical texts.

Marpa (1012-1097). An important Tibetan translator of Indian and Nepalese texts and founder of the Kagyu (*bka' brgyud*) school, who brought the Guhyasamāja Tantra and other important tantric teachings to Tibet. Marpa's disciple Milarepa became the most famous yogi in Tibetan history *mental continuum.* See *mindstream.*

Milarepa (1040-1123). One of the most revered figures in Tibetan Buddhism, in his youth he was an evil sorcerer who killed many people, before becoming a disciple of Marpa. After many years of hardship and solitary training to purify the *karma* of his evil deeds, Milarepa reached enlightenment and became a famous teacher. His spontaneous poems preserved in the *Hundred Thousand Songs of Milarepa* and his biography are among the most popular works in Tibetan literature.

mind-only doctrine. See *Yogācāra.*

mindstream (Skt. *santāna*; Tib. *rgyud*). The mental continuum of the causally connected flow of momentary instances of consciousness. It is this "stream" of mental moments, each one producing the next, that continues through the process of death, intermediate state, and rebirth.

Nāgārjuna (second century C.E.). One of the most important Mahāyāna Buddhist philosophers of India, Nāgārjuna is said to have retrieved the *Perfection of Wisdom Scriptures* from the land of the *nāgas*, and later systematized their teaching as the Middle Way or Madhyamaka philosophy. Nāgārjuna is also the name of an important tantric author.

Nāropa. A famous scholar of Nālandā monastery in northern India, he sacrificed his career and reputation to become the disciple of the mendicant yogi Tilopa. After subjecting him to twelve years of hardship, Tilopa transmitted the realization of Mahāmudrā, or "Great Seal" to Nāropa with a slap of his sandal. Later Nāropa would become the guru of Marpa, and thus the source of many of the most important Highest Yoga Tantric practices transmitted to Tibet.

natural clear light. See *clear light.*

natural pristine awareness (Tib. *rang bzhin kyi rig pa*). In the Dzogchen and Mahāmudrā systems, this is the nature of the ordinary mind. When natural pristine awareness is recognized, all appearances of *saṃsāra* and *nirvāṇa* are known as its display.

neuron. Neurons are the fundamental signaling units of the nervous system. The typical neuron transmits electrochemical information to other neurons via the axon and receives information through fibers known as dendrites.

New Translation lineage. The traditions of Sūtrayāna and Tantrayāna propagated in Tibet after the tenth century, as distinct from the Old Translation lineage, established in Tibet in the eighth and ninth centuries. Drawn from diverse Indian sources which in some cases overlap, these new lineages came to be known as the Sakya (founded by the translator Drogmi); the Kagyu (founded by Marpa); and the Kadampa (founded by Indian master Atīśa), known in its later revival by

Tsongkhapa as the Gelugpa.

Nirmāṇakāya. See *three kāyas.*

nirvāṇa (Tib. *mya ngan las 'das pa*). Literally in its Tibetan rendition meaning "passed beyond all pain and sorrow," *nirvāṇa* refers to a radical freedom from suffering and its underlying causes. Such freedom can be attained only when all negative mental states, afflictive obstructions, and the obstruction to knowledge have ceased to function. Therefore, *nirvāṇa* is sometimes referred to as cessation (*nirodha*) or release (*mokṣa*).

Nyingma order (Tib. *rnying ma pa*). Literally, the "ancient ones," the Nyingma order is the oldest school of Tibetan Buddhism. Founded in the late eighth century by the meditation master Padmasambhava, the special teachings of the Nyingma are known as Dzogchen, or the Great Perfection.

obscure phenomena (Skt. *viprakṛṣṭadharma*; Tib. *lkog gyur gyi chos*). One of three classifications of things (*dharma*) in the Indian and Tibetan Buddhist epistemological tradition. Obscured phenomena are things whose existence one knows through a valid inference, but not through direct perception. For example, one validly infers the presence of fire inside a house when one sees dense, black smoke billowing from the windows.

obstruction to knowledge (Skt. *jñeyāvaraṇa*; Tib. *shes bya'i sgrib pa*). The fundamental ignorance that underlies all of the suffering of *saṃsāra* and prevents one from understanding the identitylessness of all phenomena. Once this obstruction is eliminated, along with the afflictive obstruction, one is considered to have attained omniscience.

occipital cortex/lobe. The most posterior region of the cerebral cortex involved in the processing of visual sensory information.

Padmasambhava. The Indian meditation master who was instrumental in establishing Buddhism in Tibet in the eighth century. He is best known for his subjugation of the country's spirits and demons through his tremendous magical powers, and for propagating the teachings of the Vajrayāna.

parietal lobe. The area of the cerebral cortex that lies between the frontal lobe anteriorally and the occipital cortex posteriorally, and which is dorsal to (above) the temporal lobe. Among its most noted functions are somatosensory and visuospatial processing.

pervasive energy. One of the principal energies or "winds" of the body, it is present throughout the physical form in equal measure.

positron emission tomography (PET). A brain imaging technique capable of producing 3-D pictures of on-line brain processing. It involves the injection of radioactive substances that serve as tracers of cerebral activity.

powa. See *transfer of consciousness.*

prāṇa (Tib. *rlung*). A Sanskrit term that literally means "wind" or "breath," *prāṇa* refers to the various types of subtle energy that animate and pervade the psychophysical system. In Buddhist tantra, these "winds" or "vital energies" are brought under the meditator's control through the practices of the stage of generation and the stage of completion. The most subtle form of vital energy is identical with the most subtle form of the mind itself, and a major focus of tantric practice is the attempt to gain mastery of this most subtle energy so as to transform the mind at its most subtle level.

primary cortices. The areas of the cerebral cortex that process unimodal sensory information (i.e., information specific to a single sensory modality such as vision, hearing, touch, etc.).

primordial clear light. See *clear light.*

primordial consciousness. See *very subtle mind.*

pristine awareness (Tib. *rig pa*). Consciousness in its natural state, unmodified by conceptual constructs, by hopes and fears, by affirmation and negation. Beyond all dualities, pristine awareness is identical to the Dharmakāya, the Buddha mind.

Prāsaṅgika Madhyamaka school. A subdivision of the Madhyamaka school of Indian Mahāyāna Buddhist thought. The Prāsaṅgika Madhyamaka school is associated with Buddhapālita and Candrakīrti. Most Tibetan scholars maintain that it represents the most accurate interpretation of the Buddha's teachings on identitylessness. Unlike the Svātantrika Madhyamaka school, the Prāsaṅgika Madhyamaka school does not accept independent syllogistic reasoning, since the meanings of the terms in an argument always depend upon the interpreter of those terms. Thus, all arguments that assume the existence of an intrinsic identity can be reduced to an unacceptable or absurd consequence (*prasaṅga*).

REM sleep. Stage of sleep characterized by rapid eye movements and desynchronized EEG recordings. It is the sleep pattern most often associated with dreaming.

remote phenomena. See *obscure phenomena.*

rūpakāya (Tib. *gzugs kyi sku*). Literally the "form body" of a buddha, the rūpakāya includes both the Sambhogakāya and the Nirmāṇakāya, the bodies that are perceived by advanced spiritual practitioners and by ordinary sentient beings, respectively. See *three kāyas.*

sādhana (Tib. *sgrub thabs*). A spiritual practice of any kind. In the tantric context, the term *sādhana* usually refers to a ritual text and to the meditative techniques found within the text. In many instances, a tantric *sādhana* will include the visualization techniques of deity yoga.

Sambhogakāya. See *three kāyas.*

samādhi (Tib. *ting nge 'dzin*). A state of deep meditation where the mind is able to profoundly fathom the subject of its focus.

saṃsāra (Tib. *srid pa'i 'khor lo*). The cycle of conditioned existence in which all sentient beings perpetually revolve without choice due to the force of the *karma* and negative mental states. Also designated as cyclic existence, *saṃsāra* is the state of unenlightened or unawakened existence in which one continually encounters suffering.

Sautrāntika system. One of four major philosophical systems of ancient Indian Buddhist thought. According to the Sautrāntika system, reality ultimately consists of irreducible entities that have neither spatial nor temporal extension. These "particulars" are either material, in which case they are partless, momentary particles, or mental, in which case they are instantaneous mental moments. Although the Sautrāntika system does not, therefore, maintain the identitylessness of all things, it does assert the identitylessness of the self or person.

seizure. An initial focus of abnormal neural discharge (typically deep in the temporal lobe limbic system) that suddenly recruits a large mass of adjacent cortical tissue. The discharge is characterized as being of low frequency and is tightly synchronized.

selflessness. See *identitylessness.*

skandhas. See *five aggregates.*

stage of completion (Skt. *niṣpannakrama*; Tib. *rdzogs rim*). The second and final stage of practice in Highest Yoga Tantra, where the yogi or yogini gradually realizes the actual state of buddhahood free of contrivance, through yogic practices involving the subtle nerve channels, energy, and essences of the body.

stage of generation (Skt. *utpattikrama*; Tib. *bskyed rim*). The first stage of practice in Highest Yoga Tantra, where the yogi or yogini progressively develops a clear visual perception and identity of him or herself as a buddha-deity through various *sādhanas* and extensive mantra recitation. See also *chosen deity*, *stage of completion.*

substantial cause (Skt. *upādānahetu*; Tib. *nyer len gyi rgyu*). The primary cause of a particular effect; for example, although many conditions must be present for a green sprout to arise, the *substantial cause* is a seed.

subtle body. The network of subtle nerve channels in the body as understood in Highest Yoga Tantra, as well as the energies and subtle essences which move in those channels.

subtle channels. The subtle network which conducts the subtle energy and essences of the subtle body. There are said to be 72,000 subtle channels in the human body.

subtle clear light. See *clear light.*

subtle consciousness. See *subtle energy-mind.*

subtle energy-mind. The most subtle constituent of the subtle body, it refers to both the subtle vital energy and the subtle mind. It also refers to the *clear light,* and it is the aspect of the mind-body continuum which travels uninterrupted from one life to the next. Some say that it abides as a small sphere in the heart-center.

subtle mind. See *subtle energy-mind.*

sūtra. The exoteric teachings of the Buddha as preserved in Sanskrit and other languages. The term *sūtra,* when it is used in conjunction with the term *tantra,* can also refer to the entire system of Buddhist philosophy and practice with the exception of the secret teachings of the Vajrayāna.

Sūtrayāna. Literally, the "Sūtra Vehicle," or the path to awakening that relies upon the philosophical, ethical, and meditative systems belonging to the Sūtras, or esoteric discourses of the Buddha. By following the practices of Sūtrayāna, it is possible to reach personal liberation (arhatship) within one lifetime, or full enlightenment (buddhahood) within three "countless" eons. See also *Tantric Buddhism, Vajrayāna.*

Svātantrika Madhyamaka. A subdivision of the Madhyamaka school of Indian Mahāyāna Buddhist thought, the Svātantrika Madhyamaka school is most closely associated with Bhāvaviveka. A prominent doctrine of this school is its acceptance of independent (*svatantra*) syllogistic reasoning, in which the meanings of the terms in a syllogism are conventionally independent of the interpreter of that syllogism. Thus, things possess intrinsic identities on the conventional level.

synapses. The specialized site of communication between two neurons. Synapses can be categorized as chemical or electrical, depending on the mechanism of signal transmission and/or the properties of the synapse.

tantra (Tib. *rgyud*). Literally, in Sanskrit, meaning *continuum* or *thread,* the term tantra refers to the esoteric teachings and practices of the Buddha, as preserved in Sanskrit and other languages.

Tantric Buddhism. The path to awakening that relies upon the esoteric, or tantric, teachings of the Buddha; also known as the Mantrayāna, Tantrayāna, and Vajrayāna. There are many aspects and levels of Tantric Buddhism, but they all share in common the idea of controlling the bodily winds, or vital energies, as a means of reproducing the death process such that one can meditate on identitylessness with the most subtle level of mind. Tantric Buddhism is perhaps best known for its employment of the techniques of deity yoga, but this is only one of its many components. See also *deity yoga, prāṇa,* and *Vajrayāna.*

temporal lobe. The region of the brain anterior to the occipital lobe and ventral to (beneath) the parietal lobe. Its cortical functions include higher-order visual processing and attentional selection of visual information. Subcortical functions include aspects of learning, memory, and emotional processing.

thalamus. A midbrain structure consisting of a collection of highly specific subdivisions, or nuclei. Its connections are reciprocal; incoming sensory information is filtered and distributed to the proper primary cortical areas for further processing (bottom-up), while higher-order cortical processing, via the thalamus, influences the selection and perception of sensory data (top-down).

three kāyas (Skt. *trikāya*; Tib. *sku gsum*). The doctrine of the three *kāyas*, or bodies, presents the Mahāyāna understanding of the nature of perfect enlightenment, or buddhahood. The Dharmakāya, or Reality Body, is the ultimate expanse that is the final reality of a buddha's awakening; it is also the ultimate mind of a buddha. The Sambhogakāya, or Enjoyment Body, is the form of the enlightened mind that remains in the perfected realms of existence. This subtle form is only perceptible to highly advanced spiritual practitioners. The Nirmāṇakāya, or Emanation Body, is the form of the Buddha that is perceptible to ordinary sentient beings like ourselves.

three poisons. The basic negative mental states of desire, aversion, and ignorance.

transfer of consciousness (Tib. *pho ba*, pronounced *powa*). A method used at the moment of death to direct one's own or another person's consciousness to a happy rebirth, most typically in the Sukhāvatī paradise of the Buddha of Limitless Light, Amitābha.

transmitter. A chemical substance that is released by the axon terminal of a neuron into a synapse. The transmitter travels across the synapse to bind with a chemical receptor located on a dendrite or the cell body of the postsynaptic neuron.

Tsongkhapa (1357-1419). Founder of the Gelug tradition of Tibetan Buddhism, he revived the tradition of the Kadampas (see *New Translation lineages*) and studied, practiced, wrote, and taught extensively on the Sūtrayāna and Vajrayāna systems of all the New Translation traditions. Tsongkhapa's interpretation of the Madhyamaka or Middle Way philosophy, as preserved in his monumental commentaries on the works of Nāgārjuna and Candrakīrti, has exerted immense influence on the development of all schools of Tibetan philosophy.

Vaibhāṣika. One of the four major philosophical systems of ancient Indian Buddhist thought. The Vaibhāṣika system maintains that the universe is composed of a limited set of irreducible elements called *dharmas.* All compounded things can be reduced to these *dharmas,* which themselves possess an unchanging identity in the past, present, and future. Although the Vaibhāṣika system does not, therefore, maintain the identitylessness of all things, it does assert the identitylessness of the self or person.

Vajrayāna. Literally in Sanskrit the "Vehicle of Indestructible Reality," the Vajrayāna is the name for the path of Tantric Buddhism. In this context, the term *vajra* refers to the indestructible reality of the natural identitylessness of all phenomena, which is utilized as a vehicle to awakening. For example, the body is generated as the body of a deity, and the speech is transformed into enlightened speech, through visualization and mantra. See also *Tantric Buddhism.*

vase breathing/vase meditation. A yogic exercise where the diaphragm is contracted while the breath is held to form the shape of a vase, to expedite the process of the *inner heat* cultivated by meditators on the stage of completion.

very subtle consciousness. See *very subtle mind.*

very subtle mind. In the Vajrayāna, the very subtle mind is the *clear light.* By learning how to control the vital energy, one learns how to recognize the very subtle mind and to use it to meditate on the ultimate identitylessness of reality.

very subtle mind-energy. The continuum of the very subtle mind and the very subtle vital energy that is the most subtle basis for the designation of the self according to the Vajrayāna.

vestibular system of body balance. Sense organ (semi-circular canals) located in the inner ear that is involved in the maintenance of balance and equilibrium.

vital energy. See *prāṇa.*

wind, bile, phlegm. The three humors of the body according to Ayurvedic and Tibetan medicine. Bile is associated with heat and the emotion of anger, phlegm with cold and the emotion of ignorance, and wind with the emotion of desire.

yoga. In Sanskrit, yoga literally means "union." Yoga is any method for achieving union with the state of enlightenment.

Yogācāra. A school of Indian Mahāyāna Buddhist philosophy emphasizing the phenomenology of meditation. The Yogācāra philosophers analyze the mind into eight consciousnesses—five sensory, and three mental—including the foundation consciousness. This school is known for its idealistic doctrine of "mind-only" (*cittamātra*), in which all phenomenological appearances are understood to arise from the imprints placed in one's mindstream through the force of *karma.*

yogi/yogini. Literally, a male/female practitioner of yoga; in the Tibetan context, the terms refer to adepts of tantric meditation.

Contributors

JEROME ("PETE") ENGEL JR.

Pete Engel received his M.D. in 1965 and a Ph.D. in physiology in 1966, both from Stanford University. Since then, he has been associate and then full professor of neurology at UCLA Medical School. He has been active in a number of professional societies, serving as president of the American Epilepsy Society and the American EEG Society. He is also the editor of several professional journals including *Advanced Neurobiology*, *Epilepsy*, and *Journal of Clinical Neurophysiology.*

Dr. Engel is the editor of several volumes on epilepsy and clinical neuroscience, and recently authored *Seizures and Epilepsy* (Philadelphia: F. A. Davis, 1989). He has contributed over 110 papers in professional journals such as *Epilepsy Research*, *Journal of Neurosurgery*, *Neurology*, *Electroencephalography Clinical Neurophysiology*, and *Annals of Neurology.*

Dr. Engel brings thorough experience with states that arise during and following epileptic seizures, diabetic coma and starvation, the hallucinations that occur in conjunction with brain lesions.

Readings circulated for Mind and Life IV:

- Engel, J. 1990. "Functional Explorations of the Human Epileptic Brain and Their Therapeutic Implications," in *Electroencephalography Clinical Neurophysiology* 76: 296-316.

- Engel, J. 1990, 1991. "Neurobiological Evidence for Epilepsy-Induced Interictal Disturbances," in *Advances in Neurology* 55: 97-111.

JAYNE GACKENBACH

Jayne Gackenbach received a Ph.D. in experimental psychology from Virginia Commonwealth University in 1978. She then spent over a decade as assistant and then associate professor, mostly at the Department of Psychology, University of Northern Iowa. Currently, she works independently in Edmonton, Canada, and is a leading member of the Association for the Study of Dreams and the Lucidity Association.

Dr. Gackenbach has edited several books including *Conscious Mind, Sleeping Brain: Perspectives on Lucid Dreaming* (New York: Plenum Press, 1988) and the forthcoming *Higher States of Consciousness* (New York: Plenum Press), and has written the popular *Control Your Dreams* (New York: Harper-Collins, 1990). She is the author of several dozen articles in professional journals such as *Journal of Social Psychology, Lucidity Letter, Journal of Mental Imagery*, and *Sleep Research.*

She has distinguished herself in all aspects of lucidity research, including physiological, psychological, and transpersonal dimensions. This topic has also been of great interest to the Tibetan tradition.

Readings circulated for Mind and Life IV:

• Gackenbach, J. 1991. "Frameworks for Understanding Lucid Dreaming: A Review," in *Dreaming* 1: 109-28.

• Gackenbach, J. 1991. "A Developmental Model of Consciousness in Sleep," in *Dream Images: A Call to Mental Arms*, edited by J. Gackenbach and A. Sheikh (New York: Baywood Publishing Company, 1991).

JOAN HALIFAX

Joan Halifax received a Ph.D. in medical anthropology/psychology from University of Miami, 1968. Since then, she has held diverse positions, including researcher in Ethnomusicology, Columbia University, NIMH, and head of the Ojai Foundation, CA. Currently, she is president of the Upaya Foundation,in New Mexico, which includes a community facility for the dying.

Dr. Halifx is the author of several articles and books including *The Human Encounter with Death* (with S. Grof) (Norton, 1973), *Shamanism* (Cross Roads, 1984), and *Fruitful Darkness* (Harper and Row, 1994).

She has carried out extensive cross-cultural studies of various topics and pioneered studies on death and dying. She is also a Buddhist practitioner and a lineage holder in the Tiep Order of Thich Nhat Han.

THUPTEN JINPA

Thupten Jinpa received his monastic training at Zongkar Choede Monastery and at Gaden Monastic University in India, leading to his degree as Lharam Geshe in 1989, the Tibetan equivalent to a doctorate in divinity. Since 1986 he has been the principal translator to His Holiness The Dalai Lama on philosophy and religion. In 1989 Jinpa joined Kings College to study Western philosophy, and received his honors in 1992.

His published works include translations of the Dalai Lama's texts on Buddhist thought and practice, and papers on topics such as Buddhist perspectives on the nature of philosophy, a comparative study of Nietzschean perspectivism and the philosophy of emptiness, and the role of subjectivity in Tibetan Vajrayāna art. Jinpa brings his solid background in Tibetan and Western traditions and his extensive linguistic skills as the Dalai Lama's translator. Along with Alan Wallace, he has been the translator for all previous Mind and Life meetings.

JOYCE MCDOUGALL

Joyce McDougall received a D.Ed. from Otago University, New Zealand. Dr. McDougall was trained in psychoanalysis in London and Paris. Since 1954 she has lived and practiced in Paris, where she is now supervising and training analyst for the Paris Society and the Institute of Psychoanalysis.

Dr. McDougall is a frequent contributor to psychoanalytic books and journals in European languages and the author of several books, including *Plea for a Measure of Abnormality* (New York: I.U.P., 1980); *Theaters of the Mind* (New York: Basic Books, 1985); and *Theaters of the Body: A Psychoanalytic Approach to Psychosomatic Illness* (New York: W. W. Norton, 1989). All have been translated into many languages.

Psychoanalysis is the only Western tradition that uses a hands-on pragmatic approach to explore human experience. The role of dreams has been recognized as crucial since its very inception. Dr. McDougall is one of the most eloquent representatives of this tradition, with extensive clinical experience and an articulate theoretical understanding.

Reading circulated for Mind and Life IV:

- McDougall, J. 1990. *Theaters of the Mind* (New York: Brunner Mazel).

CHARLES TAYLOR

Charles Taylor received a Ph.D. in philosophy from Oxford in 1961. Since then he has served as assistant, associate, and full professor at McGill University, with various additional appointments, including at École Normale Superieur (Paris), Princeton University, Oxford University, and the University of California at Berkeley.

Dr. Taylor is the author of several well-known books, such as *The Explanation of Behaviour* (London: Routledge and Kegan Paul, 1964); *Hegel* (Cambridge: Cambridge University Press, 1975); and *Sources of the Self* (Cambridge, Mass.: Harvard University Press, 1989).

He is also a frequent contributor to several philosophical journals. Dr. Taylor has been an eminent contributor to both continental and Anglo-American schools of modern thought. In this meeting, it was important to have a clear understanding of current Western ideas about self, mind, and society.

Reading circulated for Mind and Life IV:

- Taylor, Charles. 1989. *Sources of the Self: The Making of the Modern Identity* (Cambridge, Mass.: Harvard University Press, 1989).

FRANCISCO J. VARELA

Francisco Varela received a Ph.D. in biology from Harvard University in 1970. Since then, he has taught and conducted research in various universities, including University of Colorado, Boulder; New York University; University of Chile; and Max Planck Institute for Brain Research (Germany). He is currently the director of research at the Centre Nationale de la Recherche Scientifique in Paris.

Dr. Varela is the author of over 150 articles on neuroscience and cognitive science in scientific periodicals such as *Journal of Cell Biology, Journal of Theoretical Biology, Perception, Vision Research, Human Brain Mapping, Biological Cybernetics, Philosophy of Science, Proceedings of the National Academy of Science (USA)* and *Nature*. He is the author of ten books, the most recent being *The Embodied Mind* (Cambridge, Mass.: MIT Press, 1991), which has been translated into eight languages.

He has been interested for a long time in the interface between Western science and Buddhism. He was the scientific coordinator of Mind and Life I in 1987, and a participant in Mind and Life III. Apart from contributing his general expertise in neuroscience, his role was largely as a moderator.

Reading circulated for Mind and Life IV:

- Varela, F., E. Thompson, and E. Rosch. 1991. *The Embodied Mind: Cognitive Science and Human Experience* (Cambridge, Mass.: MIT Press).

B. ALAN WALLACE

Alan Wallace received a Ph.D. in religious studies at Stanford University, and a B.A. from Amherst College in 1985 in physics and philosophy. From 1971 to 1979 he studied the Tibetan Buddhist tradition intensively in Dharamsala and Switzerland.

Dr. Wallace is the author of various articles on epistemology of science and religion; author of *Choosing Reality: A Contemplative View of Physics and the Mind* (Boston: Shambhala Publications, 1989). He is also the translator/commentator of several Tibetan texts, such as *Transcendent Wisdom: A Commentary on the Ninth Chapter of Shantideva's "Guide to the Bodhisattva Way of Life,"* by His Holiness the Dalai Lama (Ithaca, NY: Snow Lion Publications, 1988).

He brought an unusually deep knowledge of the Tibetan tradition combined with a broad training in Western science and philosophy to this conference. He has been the translator and counselor for all Mind and Life conferences.

Reading circulated for Mind and Life IV:

- Wallace, A. 1989. *Choosing Reality: A Contemplative View of Physics and the Mind* (Boston: Shambhala Publications, 1989).

Index